Praise for *Fresh Ba*

"Westerners, [Dr. Hernandez] writes, fall short on including Indigenous people in environmental dialogues and deny them the social and economic resources necessary to recover from 'land theft, cultural loss, and genocide' and to prepare for the future effects of climate change."

—*PUBLISHERS WEEKLY*

"In *Fresh Banana Leaves*, Jessica Hernandez weaves personal, historical, and environmental narratives to offer us a passionate and powerful call to increase our awareness and to take responsibility for caring for Mother Earth."

—EMIL' KEME (K'ICHE' MAYA NATION), member of the
Ixbalamke Junajpu Winaq' Collective

"*Fresh Banana Leaves* is a groundbreaking book that busts existing frameworks about how we think about Indigeneity, science, and environmental policy. Beyond her trenchant critiques, she also offers the generative constructs of eco-colonialism and ecological grief as new ways of thinking through the current climate crisis. This book will soon be a vanguard text in the burgeoning field of Indigenous science, a must read for practitioners and theorists alike."

—SANDY GRANDE, professor of political science and Native
American and Indigenous Studies, University of Connecticut,
and Senior Ford Fellow

"Dr. Hernandez offers many gifts for us to learn, grow, and heal. She shares many details of how settler colonialism has impacted Indigenous people, specifically people of Mexico and Central America. *Fresh Banana Leaves* is a true validation of the Indigenous knowledge of community."

—DR. MICHAEL SPENCER, Presidential Term Professor of Social
Work and director of Native Hawaiian, Pacific Islander, and Oceania
Affairs at the Indigenous Wellness Research Institute (IWRI),
University of Washington

"While ecological destruction has intensified, many of the approaches intended to minimize cataclysmic harm continue to emerge from the Global North. What has long been ignored are the practices and worldviews that Indigenous peoples have with our nonhuman relatives. *Fresh Banana Leaves* offers *seeds*—through the form of lived experiences and historic practices that come from the author's own ancestors and relatives. We are invited to take heed, to be part of rebuilding a world that is more dignified and responsive to our environment and nonhuman living relations. Our collective futures hinge upon us abiding."

—DR. ALEJANDRO VILLALPANDO, assistant professor of Latin American Studies, Cal State LA

FRESH BANANA LEAVES

HEALING INDIGENOUS LANDSCAPES THROUGH INDIGENOUS SCIENCE

JESSICA HERNANDEZ, PhD

North Atlantic Books
Huichin, unceded Ohlone land
aka Berkeley, California

Published by
North Atlantic Books
Huichin, unceded Ohlone land
aka Berkeley, California

Cover art © gettyimages.com/briddy
Cover design by Jasmine Hromjak
Book design by Happenstance Type-O-Rama

Printed in the United States of America

Fresh Banana Leaves: Healing Indigenous Landscapes through Indigenous Science is sponsored and published by North Atlantic Books, an educational nonprofit based in the unceded Ohlone land Huichin (*aka* Berkeley, CA), that collaborates with partners to develop cross-cultural perspectives, nurture holistic views of art, science, the humanities, and healing, and seed personal and global transformation by publishing work on the relationship of body, spirit, and nature.

North Atlantic Books' publications are distributed to the US trade and internationally by Penguin Random House Publisher Services. For further information, visit our website at www.northatlanticbooks.com.

Library of Congress Cataloging-in-Publication Data
Names: Hernandez, Jessica, 1990– author.
Title: Fresh banana leaves : healing indigenous landscapes through
   indigenous science / Jessica Hernandez.
Description: Berkeley, California : North Atlantic Books, [2022] | Includes
   bibliographical references.
Identifiers: LCCN 2021027787 (print) | LCCN 2021027788 (ebook) |
   ISBN 9781623176051 (trade paperback) | ISBN 9781623176068 (ebook)
Subjects: LCSH: Women and the environment—Latin America. | Indian
   women—Agriculture—Latin America. | Environmentalism—Social
   aspects—Latin America. | Environmental protection—Latin America. |
   Human ecology—Latin America. | Ecofeminism—Latin America.
Classification: LCC GE195.9 .H47 2022 (print) | LCC GE195.9 (ebook) |
   DDC 304.2082—dc23
LC record available at https://lccn.loc.gov/2021027787
LC ebook record available at https://lccn.loc.gov/2021027788

5 6 7 8 9 KPC 28 27 26 25 24 23 22

This book includes recycled material and material from well-managed forests. North Atlantic Books is committed to the protection of our environment. We print on recycled paper whenever possible and partner with printers who strive to use environmentally responsible practices.

For our Indigenous pueblos,
by our Indigenous pueblos.
May we continue to write
and tell our stories,
instead of our stories being
written and told for us.

ACKNOWLEDGMENTS

I want to acknowledge the Indigenous people who have come before me and have advocated for the protection of our Mother Earth. I want to thank my guiding ancestors for motivating me to pursue the environmental field as an Indigenous woman. I also want to thank everyone who was a part of this book and legacy: my mother, my father, my relatives, and all the Indigenous women from my communities who provided their testimonies. This book would not have been written without their support and guidance. May this book inspire you to plant seeds that will one day blossom into flowers of change. Decolonize. Revolutionize. Indigenize.

₵ONTENTS

Introduction

I am Victor Manuel. I now carry the last name Hernandez because like the war fractured my country, it fractured my identity. I had to adopt a new last name because everything was burned during the civil war, including the alcaldía, where they held all of our birth certificates and other paperwork. I was eleven years old when I fought in the war that devastated my country of El Salvador.

When I started writing this book, I asked my father to tell me more about his experiences as a child soldier during the civil war that devastated his country of El Salvador. This was the hardest and most difficult conversation I have ever had with my father, as this was something he tried to keep beneath the surface. While he shared pieces of his story with me throughout my life, he had never shared his entire story from start to finish. This was his way of protecting me from the life he endured as an Indigenous child. He never wanted his experiences to become my experiences, as his main goal as a father was to protect me from any harm. However, no matter how much he tried to hide his trauma, it always resurfaced.

War leaves a devastating impact on one's mental health, especially when the trauma is experienced during one's childhood. At a young age, I started to piece together every short story my father had shared with me to fully understand the magnitude of his lived experience. I witnessed his healing journey, which was not always easy, pleasant, or a fairy tale, as he overcame his post-traumatic stress disorder (PTSD) and other traumatic impacts the war left on him. I became his caregiver and confidant and continue to be both to him to this day. Ultimately, because I witnessed his journey, I can say that I ended up inheriting my father's trauma. This gave me a different perspective on what intergenerational trauma really means. For me, it is not just something I get to read about in health journals or hear about in research. It has manifested in my lived experience as the daughter of an Indigenous former child soldier of the Central American civil war.

My father plays an important role in my life. At a young age he had to survive on his own, and from that he developed an enriched relationship and understanding of our Mother Earth and her natural resources. Many of the things I know about our Mother Earth were lessons he taught me.

My father's knowledge, story, and lived experience have become the strongest roots of my lineage, history, and culture. Our Indigeneity has been fractured because of war and the forced displacement we had to undergo. However, these roots are allowing me to sustain my identity in the diaspora. As a displaced Indigenous woman, anywhere I end up I carry these roots with me and replant them in my new location. They are the foundation of my existence, resistance, and resilience. Who I am has shaped what I have become.

My cultural and family roots will also continue to play a major role in the legacy I will get to leave as a future ancestor. They also remind me that anywhere I go I am either an unwelcome or welcomed guest. This was a teaching my grandmother instilled in me at a young age whenever I would visit her in my maternal homelands. She would tell me, "Never forget that anywhere you go that is not your home, it is someone's home, and you must pay them respect and build relationships with the land and the people to be welcomed into their home. Otherwise, you are walking in their home as an unwelcome guest." This reminder and teaching shaped how I carry myself as a displaced Indigenous woman. I do not forget that I have the responsibility of an unwelcome guest to carry on my teachings and work to build relationships with the stewards of my new home and lands. This is a constant reminder that I am always residing on Indigenous lands, no matter if I move into a city or rural location within the United States. This entire continent of North America and the rest of the Americas—Central America, South America, and the Caribbean—are Indigenous lands. The same way I expect guests to behave in my homelands and build relationships with our land and our people is the same way I carry myself as a displaced Indigenous woman.

For non-Indigenous peoples, becoming a welcomed guest is a lifelong journey because they inherit the role of settlers. Settlers are those who are not Indigenous to the continents of the Americas, and to me, settlers do not include Black folks, who were forcefully taken from their ancestral homelands during slavery. This is the embodiment of how the United States was built. It was built on stolen lands and built by stolen Indigenous peoples from the continent of Africa. To me, Black people are included in the Indigeneity discourse and the Indigenous scholarship I write about. Black people are Indigenous; their Indigeneity was fractured over generations of the slavery, segregation, and discrimination they have endured when they were displaced from their ancestral homelands. Like Indigenous peoples, Black people have also been caretaking and stewarding our lands since slavery. This was an inherited knowledge system; they were not taught how to caretake or steward the land, as they already knew how to do this, and this is what colonizers and settlers took advantage of.

Understanding one's positionality allows one to foster the type of relationships they have to work toward and the actions they have to take as either settlers or unwelcome or welcomed guests. Everyone must reflect on what role they play in this system that the United States was built on, *settler colonialism*. Settler colonialism is the systems that continue to grant settlers the power to lead political regimes, government institutions, and natural resource allocation over the Indigenous peoples who used to coexist with the lands that are now colonized. Settler colonialism favors settlers, and thus everyone must understand whether their positionality on Indigenous lands is that of a settler, an unwelcome guest, or a welcomed guest.

3

As an Indigenous woman from Mexico and El Salvador, I navigate three distinct settler colonial systems and frameworks and will weave all three throughout the book. I navigate distinct systems because Mexico operates on a different settler colonialism than the United States and El Salvador operates on a different settler colonialism than Mexico. This difference is because *settlers* in each context refers to a different group of people, but they are ultimately people who embody *whiteness*, the white identity granted to them by each country. What positionality we embody as settlers or unwelcome or welcomed guests should be something everyone questions and self-reflects on. My grandmother ensured to teach me this, and this is definitely an important lesson I will pass down to the future generations.

As long as we (my relatives) are displaced, we have to work toward building those relationships with the lands and people. For most white people, they will forever be settlers as long as the Americas are ruled by settler colonialism. For some displaced Indigenous peoples, we will always be unwelcome guests on other Indigenous lands. Whether we are overall welcomed by the Indigenous peoples whose land we are residing on is ultimately their decision. For my home state of Oaxaca, this decision of who is a welcomed guest is up to our Indigenous pueblos. This will continue to be ingrained into our future generations until settler colonialism is dismantled.

Thinking about the seven generations has allowed me to foster a deeper understanding of how the actions I take today will impact the future after I am gone. My family and relatives have faced many life-changing events that have further shaped the way I think and the advocacy I continue to lead despite being away from my ancestral lands. My history is not what is offered in history books, but I always had a strong source among my grandparents, parents, and relatives. This is what has allowed us to carry our stories and continue to pass them to the younger generations.

My father always told me, "If I was able to survive war as a child, you will be able survive anything in life," and this teaching has fueled my resilience and resistance as an Indigenous woman. Throughout my life, I have also learned that it is our resistance and resilience that have helped us survive. However, both our resistance and resilience should not be romanticized but rather respected. Oftentimes, many stories about our Indigenous communities focus on either our vulnerability, resilience, or resistance. There is much more to Indigenous communities, but people tend to focus on these three because this helps them build an innocent narrative or spread naiveness among settlers and their descendants. If one's lineage is responsible for forcing this vulnerability, resilience, or resistance onto us, one is responsible for working hard to ensure these narratives are no longer needed. This means that settlers have to continue working hard to dismantle settler colonialism while not co-opting or stealing Indigenous knowledge, movements, or ways of life.

For many communities, settlers are within the context of the United States, Mexico, and El Salvador. Since the settlers of Latin American descent are considered people of color and oftentimes *Latinos* within the United States praxis, they tend to forget that they also have to work toward undoing settler colonialism back in our countries. They often move toward navigating their own oppression within the United States while forgetting that they are sometimes our oppressors back in our native lands. This is not to negate that settlers of Latin American descent face discrimination, racism, or xenophobia in the United States, but to remind people that we navigate various layers of settler colonialism as displaced Indigenous peoples from Latin America.

As a displaced Indigenous woman, I have to navigate both the settler colonialism that exists in the United States, Mexico, and Central America. Settlers from Latin America get a pass when they visit their homelands or return, but when I return back to my native lands, I still have to navigate a different form of settler colonialism. This is a critique against Latinidad and how it formulated a narrative of the oppressed while ignoring the narrative of the oppressor and how both play a role within Latinidad positionalities. Latinidad is a duality between the oppressed and the oppressor as settler colonialism frameworks exist in Latin America that continue to harm Indigenous, Afro-Latino, and Afro-Indigenous peoples.

Settler Colonialism and Latinidad

Growing up, I always knew that we were displaced from our native lands because of war. I knew at a young age that we did not reside on our maternal or paternal ancestral lands because my father had to leave everything behind in order to survive. He met my mother in her pueblo located in Oaxaca. Her pueblo had granted my uncle and father refuge. This was two years after being in war and being forced to escape it in order to survive. However, things were not always easy for my father as xenophobia is prevalent in Mexico.

Xenophobia and anti-immigrant policies are a domino effect throughout the Americas. For example, people in Guatemala do not like Salvadoran immigrants, people in Mexico do not like Central American immigrants, and in the United States most of the anti-immigration narratives are against Mexicans. This is how the different settler colonialism frameworks of the Americas operate. They pit each settler state against each other as their ultimate goal is to protect whiteness. Whiteness is distinct in each country, and this is why Latinidad needs to be dissected and the settler colonialism embedded in it needs to be talked about more.

The xenophobia my father experienced is what ultimately led him to seek refuge in the United States. My mother decided to support my father and followed his journey. I always knew that our residence in South Central Los Angeles was not permanent because my parents always longed to return back to their homelands. They would mention this longing to return during holidays, birthdays, and important dates as we were and continue to be the only ones from both our maternal and paternal families residing in the United States. I knew that there were people that looked like us residing in Los Angeles as well, but oftentimes, I still felt like I did not belong.

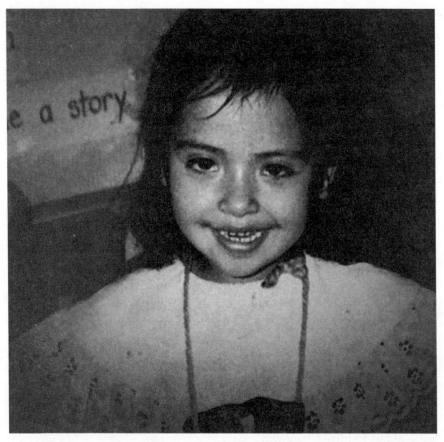

PHOTO 1.1: Young Dr. Jessica Hernandez, Family Archive. Photographer: Unknown

At a young age, one thing I did not understand was why I lacked a sense of belonging in many settings throughout Los Angeles. The only times I felt like I belonged were when I would meet another person from Oaxaca and, sometimes, another person from El Salvador. Being young, I was not fully aware of the racial systems and hierarchies that exist in Latin America because in the United States we were all and continue to be identified as *Latinos*. I always felt an uneasiness and uncertainty in many spaces throughout my life in Los Angeles. Our primary place that I frequented the most was school. In many of our school functions, I noticed that my mom would attend out of obligation and not invitation. She has always been a quiet person and never connected unless she met another adult from Oaxaca. I always thought my mother was timid, but like me, she also felt like she did not belong.

It wasn't until I grew older that I started understanding that within Latinidad, or the Latino identity, Indigeneity is not fully embodied because the United States tries to make us all monolithic despite the racial hierarchies that exist in Mexico that include white, mestizo, Indigenous, Afro-Indigenous, and Afro-Mexican. This racial hierarchy is what is sustained by settler colonialism power and privilege granted to both white Mexicans and mestizos. In the United States context, though, we are all the same, and this is why those of us who are not granted power and privilege back in Mexico feel like we do not belong in these Latino spaces. The same applies to Central American spaces. The racial hierarchies that exist there are the same except that in many Central American countries, mestizos are referred to as *ladinos*.

Latinidad is the umbrella racial term that the United States developed to make all people from Latin America be considered *Latinos* or *Hispanic*. South Central Los Angeles has a high number of Latinos, so it should be expected that I would feel seen, heard, or understood. However, as I grew older I started to grasp that while we may all share this identity label bestowed upon us as Latinos in the United States, we are all different because of the racial hierarchies and structures that exist in Latin America. Back in our countries the racial caste systems are still present, and these systems grant power and privilege to whites and mestizos. Thus Latinidad does not represent the complexity of how race plays a major role in the power and privilege we are granted back in our lands. Latin America follows a different settler colonialism framework because unlike the United States, the colonizing countries that played a major influence in Latin America were mostly Spain and Portugal. The Spanish and Portuguese monarchies and colonial firms carried the colonization of most Latin American countries, so the settler colonialism framework is different from the framework that exists in the United States. Most of the settler colonialism frameworks in existence in the United States were bestowed by mostly the British and French monarchies and colonial firms. Yes, there are other European and settler countries that made their ways and had some influence in the Americas, but these four tend to dominate.

In the United States we face this acculturation of our experiences and identities under the Latinidad framework. This framework labels every person having roots to Latin America as *Latino* or sometimes as *Hispanic*. These sets of identities monopolize our lived experiences, but given that racial hierarchies and settler colonialism still play a major role in Latin America, it divides us more than unifies us. It is no surprise that most Latino representation is that of whites or mestizos because they are closer to the whiteness systems the United States operates under. I was reminded that I was not mestiza or Latina every time we visited my mother's maternal lands of Oaxaca; I knew that our positionality was different from other Mexicans. From the time we would step foot in Mexico, we knew that my father would be asked for *mordidas*, as they call the act of police officers asking you for money, because he was not Mexican. This tends to happen a lot to Central Americans in Mexico as the police are blatantly corrupt.

I also knew that going to places with my grandmother was going to result in interesting situations where we would be called racial slurs. Being called *india* or *oaxaquita* was very common to hear when I would go to the grocery store with my grandmother back in Mexico. At such a young age, I was starting to see how the identity I was given in the United States that attempted to make me feel like I was similar to other Latinos was not the same identity I embodied when I would return back to my ancestral lands. This was because the racial hierarchies in Mexico make it very clear who is Indigenous and who is white or *mestizo*. No one attempts to identify themselves as Indigenous back home if they are white or mestizo—something that is different in the United States as oftentimes many discourses try to portray the false narrative that all Mexicans are Indigenous. This is false, and if it were true, as Indigenous peoples of Mexico and the rest of Latin America, we would not have to be continuously fighting for our rights, land, or livelihoods. This narrative tries to free white or mestizos from Latin America from the work they also have to do as they play a major role in the settler colonialism that governs our countries.

In Mexico, *mestizos* have more power and privilege than Indigenous peoples and it simply denotes someone's mixed heritage with Spanish and Indigenous. In Mexico, we also know that Spanish or any other European lineage and heritage provides one more power because it is the racial hierarchy that is favored in the country, so those of us who are Indigenous are taught that our place is at the bottom of the hierarchies. This is why you see the inequities that exist in Mexico drastically impact our Indigenous pueblos rather than those who are mestizo or white.

In relation to my father's story and role in the war, his identity and socioeconomic class are what determined that he was going to be one of the children that would be forced into such a violent and brutal war. As an Indigenous boy, he did not have the means to buy a safe ticket out of the country or any protection from having to fight in the war. He was only eleven and was forced to leave his childhood behind in order to survive the war at that young age.

The role Central Americans played in the civil war can help determine their socioeconomic status. Those who fought were poor and/or Indigenous and had no means to avoid being in war. This is something my father continues to remind me of. While it is still coined as a war, the Central American civil war was also a genocide carried out to murder Indigenous peoples. It is no surprise that the United Nations has deemed the Civil War of Guatemala a genocide because of the high number of Indigenous peoples that were murdered and has compared the Salvador civil war to other genocides that took place around the globe.[1] The same narrative was carried out in El Salvador, but our country is smaller, so the numbers were not as high as those of Guatemala. However, in both situations the war was a way to suppress Indigenous peoples who were being labeled as insurgents trying to overthrow the oppressive and settler government. Given that settler colonialism is an important framework for all the Americas, this can be another reason why the United States supported these governments and provided them with resources to carry out the genocide against Indigenous peoples, mostly from Maya communities and pueblos.

My Journey as an Indigenous Scientist

Navigating multiple frameworks of settler colonialism from Mexico, Central America, and the United States has shaped the way I view my environment and places. It has also granted me a magnifying glass that serves as my lens at dissecting settler colonialism in the environmental discourse transnationally. From conservation to restoration, I am able to culturally ground both Western perspectives to my cultural values as an Indigenous person. Despite Western education and the sciences continuing to try placing me in boxes, my worldviews are interdisciplinary like my academic journey. We are often taught that within academia, we must pick one field of study and stay within that box and system. In Western academia, if you study biology and then integrate ecology, you are considered interdisciplinary. However, to me, both biology and ecology are within the same holistic system that shapes our worldviews as Indigenous peoples.

On top of this, as an Indigenous person, I can see how both biology and ecology are interconnected to health, education, and other systems that are deemed far removed from biology or ecology within Western academic frameworks. This is due to the holistic way of thinking and knowing that we hold as Indigenous peoples, that everything is ultimately connected to us and our environments. My academic career was in the environmental sciences, ranging from marine science to forestry. However, I have incorporated education, urban ecology, environmental, food, and climate justice, and I am currently working on energy justice. While this may be deemed as scattered, this is the way we as Indigenous peoples look at the world. Everything in our environments has a relationship with us and this is why it is hard for us to box things like Western ways of knowing does.

This constant push to not be placed in a box is why I adopted the persona of an Indigenous scientist. To me, an Indigenous scientist is an Indigenous person within the sciences who integrates their ways of knowing and culture into their field. They also work toward undoing layers of settler colonialism within their fields to ensure that Indigenous cultures, and most importantly, Indigenous peoples, are respected, included, and amplified. Being an Indigenous scientist ultimately translates to the responsibilities we hold with our communities and the communities whose land we reside on. Unlike Western scientists, we navigate our fields of studies having a symbiotic relationship with Indigenous communities and move away from this extractive research process Western science teaches us to conduct. While in Western academia, we are always in competition with one another; as Indigenous scientists we collaborate because we know that we are ultimately stronger together. Being an Indigenous scientist means that our communities come first, and we will continue building frameworks and paradigms to shift settler colonialism within the sciences. As a displaced Indigenous scientist, this means that the work I have to do is transnational as I continue to hold responsibilities with my communities back in Oaxaca and El Salvador. These responsibilities are one of the major reasons why I pursued a career in the environmental sciences.

Living in the diaspora is an everyday reminder of how our Indigenous cultures continue to be impacted present-day. Unfortunately, Indigenous cultures are still under attack because of settler colonialism and the impacts it has had and continues to have on our environment. As Indigenous peoples, we continue to face the impacts of climate change. While climate change is often a discourse far removed to many, for Indigenous peoples our well-being and livelihoods are already being impacted because of climate change. Climate change impacts, coupled with the pandemic that initiated in 2020, made it harder for us to cope with both.

We bore witness to the natural disasters that impacted Indigenous communities in Latin America.

Despite climate impacts being highlighted and amplified during the pandemic of 2020, Indigenous narratives continue to be dismissed and ignored in mainstream environmentalism. The environmental discourse has failed and continues to fail in uplifting and centering Indigenous peoples' voices, perspectives, and lived experiences. Ultimately, this creates further marginalization against Indigenous peoples as our voices, perspectives, and lived experiences are crucial and necessary to incorporate. Without them, we are continuing to separate humans from nature, and this is why our environments are at the current state they are today. It is important to note that Indigenous peoples have been the stewards and caretakers of our environments since time immemorial. Yet we are often left out from the environmental discourse and any decision-making pertaining to our environment.

In order to heal our environments, which are all Indigenous lands, we must incorporate Indigenous voices, perspectives and lived experiences. In this book. I also refer to these as Indigenous science because they embody our ways of knowing that are rooted from ancestral knowledge and valid sciences. I personally do not like to use traditional ecological knowledge, even though sometimes this is the only way we can refer to our Indigenous knowledge systems, because to me, traditional ecological knowledge places us within past contexts. It is important to note that the same way our environments have adapted, our Indigenous knowledge systems have adapted, and this is why I view it as a science in itself—Indigenous science.

As an Indigenous scientist, I want this book to help facilitate discussions of the importance of not only uplifting but also centering Indigenous peoples in discussions about conservation, restoration, environmental sciences, and anything that interconnects with our environments. I want us to move away from making Indigenous-led movements hashtags into creating tangible solutions that will allow us to continue protecting our environments and sustaining our cultures. Too often we are the frontline communities of the environmental, climate, food, and energy justice movements, yet when it comes to being granted a seat at the table, the table never has a reservation for us. On the contrary, we continue to have to create our own tables, but that is not where the political power sits.

It is also important to mention that one book cannot change our environmental discourse, but if seeds can be planted here and there, those seeds will sprout and create a ripple effect that can lead to change. Through my own Indigenous ways of knowing, perspectives, and lived experience, I hope this book also allows any Indigenous community member, scholar, leader, advocate, and others to also use it as a tool to get rid of that emotional labor that comes from always having to teach people why including us at the table is crucial. Our voices are invalidated unless we can provide tangible products to validate them. This is the Western way of thinking and unfortunately, for many of us, there are many books written about us, but a few written by us. Here I share my story of my family, starting with what led to my family's displacement and share community members' testimonies that center Indigenous peoples from Mexico and Central America. I think having the opportunity to do so makes me reminisce about what my father told me when I called him to tell him I had passed my dissertation. He told me, "Now the work has begun and your responsibilities with your communities have increased as you have a powerful tool that is rarely granted to us, higher education."

My father's statement made me reflect on why I felt like I needed to pursue a doctoral degree in the environmental sciences. My credentials were something I had to seek in order to somewhat receive recognition and some validation for the ancestral knowledge I carry. This has not been an easy journey as under Western science my ancestral and Indigenous knowledge is invalidated. I have often been asked to cite my lived experiences under the guise of Western academia because oral stories are not as valuable as peer-reviewed journal articles. We, as Indigenous peoples, have often been asked to do this because, for academia, lived experiences and perspectives are not important unless they have been developed into a framework or theory. I can recall the many times I have been laughed at by professors with prestige when I have analyzed certain scenarios through my lived experiences and at times, they have asked me in a mocking way whether that was Jessica's theory.

It is a weird feeling to be in academia and constantly be invalidated for the knowledge we carry, because this knowledge is also a part of our identity. It is as though with our knowledge being invalidated, we as Indigenous peoples and scholars are also invalidated. I hope that you read this book with an open mind and while understanding that in order to truly undo what settler colonialism has created, there is a lifelong journey everyone has to take a part in. In this book, I not only focus on settler colonialism from the United States lens but also from the Mexican and Central American lens as those are the three distinct settler colonialism frameworks I navigate every day as a transnational Indigenous woman.

Thank you for deciding to learn from my communities, relatives, and me throughout this book, *Fresh Banana Leaves: Healing Indigenous Landscapes through Indigenous Science.*

1

—

Indigenous Teaching:
Nature Protects You as Long as You Protect Nature

The Story behind Banana Trees

I am often asked why I decided on the title *Fresh Banana Leaves* for this book. It was not hard for me to come up with this title because there is a strong storyline behind banana leaves. Oftentimes, we hear the phrase that our ancestors are watching over us, but my father always told me that our animal and plant relatives are also watching over us. He always told me that as long as we protect nature, nature will protect us. In order to fully understand what he means, I want to contextualize the part of his story that is the foundation for the title of this book and situate the conversation I had with my father about his experiences in the civil war of El Salvador.

The civil war in El Salvador started because there was a huge political divide among the upper and ruling classes that held all the political power. This political power further oppressed the poor and working class, including all Indigenous communities. As a result, the poor rural and working class along with Indigenous communities formed resistance groups to take action against the oppression they continued to face.

These community-led resistance groups formed in the late 1920s created fear among the Salvadoran government, the upper class, and their allies because it questioned their power and privilege that they had worked toward maintaining. Thus the government initiated a military dictatorship that would murder civilians who were accused of organizing resistance groups.[1] This violent tactic was used to try to control anyone who was considering starting or joining a resistance group.

It also aimed to get rid of existent resistance groups because the government and the upper class were willing to do anything they could to maintain their power and privilege. They also used this military dictatorship because the Salvadoran government and its allies were also afraid of communism spreading in the country and taking over the government structure. The United States was a major political and economic supporter of the Salvadoran government during this time as the United States' goal has always been to prevent communism from spreading in the Americas.[2] Thus they provided the Salvadoran government with economic and military aid. It is known that this military dictatorship (*escuadrones de la muerte*) had received weapons and military training from the United States government and military branches.[3] Therefore, the tactics they were using were new and foreign to many Salvadorans.

Given the violent tactics and the continued murders of community leaders, including Óscar Romero, a priest who was known to speak against these violent atrocities, the community had had enough. Óscar Romero was assassinated on March 24, 1980, in San Salvador, El Salvador, while giving mass.[4] His death and the continued political intimidation the Salvadoran government was using against civilians led to the creation of the *Frente Farabundo Martí para la Liberación Nacional* (FMLN).[5] This party was named after Farabundo Martí, a community leader who was murdered during *La Matanza* (a military-led massacre against peasants and Indigenous communities, mostly Nahua people) on January 22, 1932.[6] This massacre was led by the government due to the social unrest and fear that were growing in the country.

What spurred this *matanza,* massacre, was that peasants and Indigenous communities were tired of the oppression they were subjected to by large corporations that were allowed to do land grabs in El Salvador. Land grabs are large-scale land purchases and acquisitions that favor large agricultural corporations, from coffee plantations to cattle ranches. During the 1920s coffee plantations had taken over El Salvador. Presidents during this time period included both Pio Romero Bosque and Arturo Araujo,[7] who were the main leaders behind taking land from the Indigenous communities and privatizing them so they could sell them to be used for coffee plantations. Many of these coffee plantation landowners employed Indigenous communities, including the Nahua people, and paid them through food, not a monetary stipend. They provided them with beans and tortillas, which are not an expensive food source, so the landowners' profits were extremely high since they had "cheap labor." Indigenous communities during these times not only had their lands, through land grabs, stolen but they were also being exploited for their labor by the large plantations that purchased their lands. As a result, the Nahua community wanted to revolt against this oppression. They organized to oppose this oppression, and this is what led to the *gran matanza.*

It is important that while the Indigenous communities and peasants had machetes as their only form of weapons, the Salvadoran government had access to expensive military weapons that were granted to them by the United States. The Salvadoran military marched through the areas where the coffee plantations were located and murdered any Indigenous person they came in contact with. They targeted people based on their clothing and the language they spoke, mostly Nawat. As a result, the Nahua community stopped wearing their traditional clothing, and because of the persecution they faced during the 1920s, coupled with over forty thousand Nahua people and allies murdered during this time, the Nawat language also became endangered.[8]

This massacre is one of the many forms of oppression we continue to face under settler governments. Settler governments want to maintain

19

their political structures; therefore, any type of uprising or resistance becomes their political enemy. If Indigenous peoples form coalitions and political organizations to stand up for their rights, they are silenced by the settler governments. This massacre ended up with taking over forty thousand lives, and it was all over coffee and its commercialization under a capitalist regime. This coffee was grown and produced on Nahua lands and territories in El Salvador. The corporations not only acquired Indigenous lands through land grabs, but also asked for and demanded labor from Indigenous peoples and only paid through beans and tortillas, food. It is as though Indigenous peoples should be happy with receiving the short end of the stick under settler governments.

Oftentimes, we as a society become naive to the oppression that governments enact, especially those of us with power and privilege. This power and privilege always favor whiteness and settlers. In 2021, the president of El Salvador Mauricio Funes recognized this massacre and apologized to the Nahua people.[9] However, apologies can only do so much when over forty thousand of your people and allies were murdered.

Since the murder of Óscar Romero and the *gran matanza*, the military continued its violent tactics until the civil war eventually broke in the late 1970s. It is important to mention that before the civil war started, over fifty years of military persecution and dictatorship had taken place. This, coupled with the social divides that had started in the country since the late 1920s, did not fulfill the government's goal of controlling or calming civilians, peasants, and Indigenous peoples who were not content with what was taking place in the country. Thus the civil war broke in the late 1970s, and this is the start of my father's (Victor's) story. He was only eleven years old when the war started in his country of El Salvador.

Dad, can you tell me more about your experience during the civil war of El Salvador?

I had eleven years when I joined to fight in the war. I volunteered, but that is because I did not have a choice. Soldiers were visiting our homes recruiting us, but mostly forcing us to join. My father had already passed when I was eleven and since his passing, I had been working to support my mother and four siblings. At eleven, I was considered old enough to join so the soldiers kept visiting my house looking for me so they could take me and force me to join the military and fight in the war.

Let me try to fully understand the reason why you joined the war. It was mostly something you were forced to do rather than a choice you made?

Yes. I joined the war because the *escuadron de la muerte* (death squads) was forcing people, mostly men and young boys to join. If you did not join they would murder you in broad daylight. They did not care who was there to witness your murder. Many children were murdered in front of their mothers. You had to find a way to save your life. Death would force you to join. I was very good at hiding in the *monte* (hills surrounded by tall grass and trees) when the *escuadron* would come looking for me. They got so tired of not finding me that they burned our little house. Fortunately, none of us, including my mother and siblings, were not at home during this. This was a common tactic they would use, and they did not care if anyone was inside the house when they would set fire to our homes. They would burn people alive. These were the forceful and violent tactics they would use to force us to join. After they burned our home, I decided to join the guerilla because this was the only way I could get protection for my family and myself. The government used these *escuadrones* to terrorize citizens, in particular poor and Indigenous peoples and instill this fear to not revolt against them. They used *escuadrones* to recruit you before the guerilla did. I hated their violent tactics and yes, we witnessed many people be burned alive because they did not want to join or let their children join the *escuadrones*. The only thing I remember adults telling me was that these *escuadrones* had received training by the United States military. So their tactics and weapons were things we had never seen before in our small country.

The same country that continues to mistreat Central American refugees, including our people from El Salvador, is responsible for the atrocities committed against civilians during the war?

Based on what the guerilla would tell us when we were training to join, yes. I became fearful of what they could do to me or my family, so after the last time escaping to *el monte*, I went in search of the guerilla because I knew I needed protection. I received training—*de masa* is what they would call it because I was not yet active in combat. But honestly, from the point I joined the guerilla, I was already traumatized and scared because I had far often witnessed how the *escuadrones de la muerte* would come to my *canton* or pueblo, whatever you want to call it, to burn homes and murder people. I think this was also a part of their tactics, because they wanted to make sure that if they did not kill you, you became afraid of them that you will not join the guerilla by no means.

Based on your lived experience in war, why do you think you survived?

I will say that I survived because nature served as a protective shield for me. It took care of me when no one would. As I mentioned, I would go hide in *el monte* when the *escuadrones de la muerte* were coming to look for me. They knew of every child or adult who was living in my *canton* because the ones who would join would tell them about the rest of us. Nature not only protected me, it saved my life. I remember I was fourteen at the time and our guerrilla encampment was bombarded because we were attacked by the military. I thought my life was going to end because I remember I sought refuge under a banana tree. This was one of the banana trees we had in our encampment. For a couple of months, it served as our only food source for the entire guerilla. We survive off bananas from these trees and tortillas we would make that we would season with salt. Now, I recall during this bombardment of our encampment that a bomb fell on top of the tree I was under. My short life flashed before my eyes, but as the bomb dropped, I saw how the banana trees wrapped its leaves around this bomb and it did not ignite. I was surprised because every bomb that was dropping was going off. I did receive gunshots during this battle. When everything stopped the guerilla took me and one of my fellow comrades, who had lost his leg, and dropped us off in the *bunque* (field). They left us with no water or food. The other comrade lost his life because he was bleeding a lot. I do not know how I managed to get myself up. I could not walk well but I managed to cut off a branch from this banana tree and used it as a crutch. I walked all the way to Guatemala, which was not far from where we were. I lived close to the border of Guatemala and our guerrilla encampment was even closer to Guatemala. Before I left I tried getting help from my family in El Salvador, but they were all scared that the military would come and murder them if they helped me as a guerilla soldier. When I made it to Guatemala an Indigenous pueblo helped me stop some of the bleeding from my leg. When I got better, I walked myself back to El Salvador to pick up my younger brother, your uncle, because he was reaching the age to be recruited by the military. Thinking back to the day when our encampment was bombarded, I strongly believe that it was this banana tree that saved my life. It is ironic because banana trees are not

native to El Salvador. They were introduced by large agricultural companies to our lands to create plantations. However, there were banana trees around our entire country. This
banana tree not only saved my life, its relatives also provided
us with food. They all nourished us as well.

My father's story is only one story of the hundreds to thousands of
children who were forced to fight in the civil war of El Salvador. However,
there is an underlying message in his story, a message that is grounded on
his Indigenous principles as a Maya Ch'orti' man: *nature protects us as long
as we protect nature.*

Why Banana Leaves?

My father not only used nature (banana trees) as his shield to protect
him from war, he used it as his food source during this time period. He
is right; banana trees are not native to our El Salvador. Banana trees were
introduced during colonialism between the sixteenth and nineteenth centuries in Latin America through the development of plantations. Banana
trees are indeed native to the South Pacific and Southeast Asian regions.
Yes, under the lens of Western environmentalism, banana trees are an
invasive species to my ancestral native lands. However, to us, bananas are
not invasive; they are displaced relatives that have adapted well to our
climates and are now incorporated into our traditional diets. Ultimately,
the kinships and relationships we have developed with them have made
them our relatives as well. All I can think of is that, like me and many
Indigenous peoples in the diaspora, banana trees have also been displaced.
We have been displaced from our native and ancestral lands and forced to
adapt to our new environments and form new kinships with our new land.

However, the juxtaposition of banana trees is that they were introduced
during colonial times to El Salvador and across Central America. While
they were used as tools of colonization, given that banana plantations

provided commonwealth to colonizers' descendants,[10] they ended up being a food source for many guerilla resistance groups against the military dictatorship that was doing everything in its power to maintain these colonial structures. One of these colonial structures is capitalism. Given that communism is the antithesis of capitalism, the government was going to do everything it could, with aid from the United States and Canada, to ensure communism did not spread in or take over El Salvador. One of the common ideologies that many Indigenous communities share across the globe is that we should never take more from nature than what we need. However, this belief and ideology do not coexist in capitalism as the main goal of capitalism is to make profit and support imperialistic power nations such as the United States. Many of these imperialistic power nations played a major role in the colonization that resulted in the genocide against Indigenous peoples. For us, the Central American civil war was also another genocide carried out against Indigenous communities, as throughout Central America we were targeted first.

Thinking about the role coffee and bananas have played in El Salvador's economy always reminds me of my dad's story. A banana tree saved him, and I think this is explained through the caretaking we continue to do as Indigenous peoples toward our nature. Indigenous peoples are employed at these plantations because there is not much training that needs to be provided to our community members. We already know how to work with nature, despite these agricultural practices that were introduced to the Americas.

The way agriculture is today is not traditional to our lands. Agriculture is a branch of capitalism because it aims to grow many crops in a land plot in order to sell massively and sometimes export to wealthy nations like the United States. It contradicts our Indigenous principle of only taking what you need from nature, because everything grown under agriculture is meant to be taken and sold. Sometimes my father's teaching of nature protecting you if you protect nature is hard to apply to natural areas where agricultural practices have taken over. How can we protect nature if it is designated to become commodified because capitalism reigns over our lands now? My father found a way to protect nature, and in his case, this included nonnative plant relatives such as the banana tree.

While his story about the bomb falling on top of a banana tree and not igniting may seem surreal, this surrealism that is embedded to our environments explains the relationship that we as Indigenous peoples have with our environments. Our environments as Indigenous peoples have always been an integral part of our identities. A part that is essential to our cultural survival and understanding of how the world shapes us as individuals. For Indigenous peoples, it transcends our past history into our present-day lives. Nature in my context embodies everything that plays a role within our landscapes. This also includes nonliving things such as rocks. I was taught that even nonliving things have a spirit if they are naturally formed or created like rocks. However, under capitalism, nature is only the thing that has a natural capital, which means that it can be sold under its set economic value.

However, for Indigenous peoples, our environment is so intertwined with who we are. It carries our memories—both ancestral and those we are currently creating. Our environments and natural resources are more than natural capital or economic revenue to us. Our environment can carry our trauma—trauma that can be traced to settler colonialism, when our ancestors were murdered, displaced, and forced to live under the Western European concepts of culture, science, and capitalism that still govern our current societies—as well as our healing. Settler colonialism created racial hierarchies throughout the Americas that continue to favor whiteness. This whiteness favors those with European ancestry over those who are Indigenous and have strong roots to the continents of the Americas. For most of us, this trauma is more present in our lives because the marginalization, oppression, and persecution against Indigenous peoples has not ended.

However, in many Indigenous narratives and discourses, we often forget to mention the resilience, determination, and strength our environment transcends to us through these shared memories that have sustained our cultures despite it all. Like our environments, we have, too, survived the impacts of colonization and continue to thrive despite the harsh conditions we both continue to experience and endure—from climate change to environmental degradation.

We still live and operate under settler colonialism in our current societies. Settler colonialism has granted non-Indigenous peoples in the Americas power and privilege, especially those closer to whiteness (white people and European descendants). Settler colonialism was led due to Western religion that granted colonizers the right from their god to colonize our lands. This belief is also known as *manifest destiny*.[11] Manifest destiny reinforced to colonizers that it was their right from God to journey to the Americas and save the people who were "uncivilized" and lived on this continent. While colonization continues to be taught as a history lesson, settler colonialism, the aftermath, and structures that were created because of this colonization continue to exist throughout the Americas. As Indigenous peoples, we are not naive to our everyday reality that settler colonialism continues to rule over our ancestral lands of the Americas.

For me personally, my environments carry the trauma that my parents had to endure as Indigenous peoples who were forced to flee their homelands. I mention environments as a plural entity because I strongly believe that while we tend to contextualize our Mother Earth as one entity, there are different environments that she maintains. Thus we all have our own environment or environments that are connected to us through various relationships. These relationships or kinships can be passed down to us from birth or they can be ones we nourish, especially through our own displacements. As an Indigenous woman whose family was displaced several times transnationally, I have more than one environment that is important to me: the environment of my paternal and maternal native lands and the current environment I am living in. My paternal native lands of El Salvador carry the violent memories my father had to endure as a child soldier in the Central American civil war. In order to survive the war, after receiving gun wounds, he had to flee and seek refuge in another country. He eventually made his journey to my maternal native lands of Oaxaca, Mexico.

Despite being displaced, those memories of trauma enacted by war are still present in our lives and tend to resurface when we remember, visit, or mention our paternal homelands. This is why I say that our environments can carry our trauma and for my father, his new environment, that of displacement, is what carried his healing. His healing started when he began reclaiming his relationship with nature in the diaspora. Caring for the plants in his small garden and watering them brought him back his joy that had been stolen from him as a child because of war. He always believed that when we take care of nature, nature takes care of us. I saw that through him because as he was taking care of nature, nature was taking care of his healing from the trauma war left on him.

My father is one of my greatest teachers, and I will always carry his teachings and stories and hope that they are passed down in our lineage. There is also a story he shared with me that puts forward one of his greatest teachings he passed down to me. This is the story that passed down his intergenerational love for nature and our environments that has been present throughout our ancestral history as Indigenous peoples. This is also what ultimately led me to pursue the environmental sciences as I was given the opportunity to pursue higher education. My deep love and respect for our environments and nature are why I became an Indigenous scientist and why I advocate that Indigenous science can ultimately heal our Indigenous lands.

His story where his life was saved by a banana tree, in particular its leaves wrapping the bomb, is why I selected this title, *Fresh Banana Leaves*, for this book. The banana leaves saved my father's life and it offered him a *fresh* start in the diaspora. His story reminds me of the harsh realities that wars continue to leave behind on our environments and our memories. He recalls hearing loud airplane noises that eventually muted all of nature's sounds. As he looked up to see what had muted his entire surroundings, he saw military planes. In an eyeblink, bombs started falling on the encampment and obliterating everything they came in contact with. The gruesome scenes that these airstrikes left were not new scenes to him as this was the same destruction bombings and military raids had left in many villages and *pueblitos* throughout El Salvador. His survival instincts told him to start running toward the banana tree while trying to avoid the shooting from the military soldiers on the ground. All he could think of during this time was to seek refugee from the bombings, and the banana tree was what called him. The same tree my father had shared his stories to and the same banana tree he had built a deep connection with was the same tree calling him toward it. This banana tree was special to him because it had become his nonhuman friend, and this friend allowed him to escape the human world that, to my father, was full of torment and violence.

My father told me that sometimes he would climb the tree to get some bananas for the rest of his comrades as most of them were afraid of heights. While on the tree, he would sit on one of its branches and just enjoy some bananas as he stared into the landscape and prayed for safety. It was the same banana tree in his recollection of events that was the determining factor to escape and leave everything behind in his native lands. At fourteen, my father had already spent most of his childhood in survival mode and had witnessed the worst that comes from human kind—war. This is why he always sought connections and relationships with plants and animals. Plants and animals are innocent and they never want to hurt us, he always told me. The more we take care of them, the more they take care of us. They communicate to each other, and they know which humans are here to not just extract from them but also respect and care for them as well.

My father saw a bomb drop above him. As he recalls, he thought his life was going to come to an end and all he could think of was "it was going to end too soon." His short years of his life flashed before his eyes. However, when the bomb dropped on top of the banana tree, the tree's leaves wrapped themselves around the bomb, preventing it from igniting. It is a surreal vision that is still hard for him to believe, but that banana tree saved his life and ultimately saved our lineage's and descendants' lives. Yes, our ancestors look after us, but sometimes they also offer us their protection through our animal and plant relatives. After the bombing stopped, everything in sight was destroyed. However, my father was standing underneath this banana tree.

There is a magical surrealism in this story but that is because the connection we as Indigenous peoples have with nature is far greater than the Western way of thinking can ever explain. This is the spiritual connection that makes us mourn when our environment is destroyed as parts of our spirits are also destroyed with our environment. The banana tree was not destroyed during the bombardment, and it became why my father survived and why I am here.

He did receive gun wounds in his leg and was left in the fields by his fellow guerrilla soldiers because, injured, he was a liability to them. As he recalled his story, he did wish they would have left at least some water for him, but they did not. He does understand why this happened as during war, in order to survive one must adopt an individualistic mentality. This means that you only look out for yourself and not others. Your main goal as a child in war is to survive as many of them had already lost everything from their families to their homes.

Taking care of nature, and nature taking care of us in return, is the greatest teaching my father has taught me. Indeed, nature protects us as long as we protect nature. This is something Western science has failed to understand or explain. Settler colonialism introduced ideologies and beliefs that nature is meant to provide us resources, to meet our needs, without requiring us to protect it as well. Nature has been described as an infinite sink, and this is what has led to overfishing, overharvesting, and essentially environmental degradation. Environmental degradation is the destruction that continues to occur in our environments. It is why our environment continues to face severe droughts, wildfires, and other natural disasters and our ecosystems continue to decline.

Indigenous peoples know that Western way of thought has also taught us that we are separate from nature. Nature is viewed for its value, whether it is economic, or for its beauty, a beauty that is rooted in the Western notion of pristine wilderness. However, both values that we continue to place on nature and its resources continue to separate us from it. These economic values (natural capital) are the main reasons why coffee and banana plantations were introduced in El Salvador. Capitalism, coupled with natural capital, is also what pushed these landowners to further exploit and oppress Indigenous peoples to generate more profits. All of these things combined created these social divides that pushed poor, working class, and Indigenous peoples to start organizing to revolt and fight for their rights.

A lot of our rainforest in El Salvador was wiped out to make room and clear the landscapes for these plantations. Many of our elders say that if the rainforest were not cleared as much as it had been, that even the current natural disasters such as hurricanes that El Salvador experienced in 2020 would not have been as terrible because the three would have reduced the flooding by capturing some of the water from the heavy rains. In Western science, this has been proven as it is estimated that trees do not only provide canopies or covers from rain, but their tree trunks can absorb up to 35 percent of rainwater. This is the estimate for mature trees, which are the ones that are fully grown and have reached their maximum height. Younger trees also absorb rainwater but at lower percentages. This sole understanding that is rooted in Indigenous understanding of our environments and the fact that our ways of knowing are ignored under settler colonialism is why we have devastated nature to the point that it can no longer protect us from natural disasters. If there were more trees in El Salvador, the heavy rains from the hurricanes would decrease in magnitude as these trees would have already been mature so they would have absorbed 35 percent of rainwater. This means that the flooding that impacted our communities the most would have decreased.

Unfortunately, we continue to value nature's capital and economic revenue instead of the protection it can grant us. If our landscapes were not degraded to make space for extractive agricultural practices, or huge cities, our landscapes would be protecting us from any climatic changes the earth underwent. However, we must learn to adapt to these new environments while ensuring to replenish what we had destroyed in nature. As long as we continue to remove ourselves from nature, nature will not be able to protect us from environmental impacts. Of course, with these environmental impacts, the most marginalized communities, including Indigenous communities, are the ones most disproportionately impacted. Settler colonialism has taught us that in the species hierarchy we are on top and therefore we have more power than other plant and animal species. It is this notion of hierarchy that favors white people in the racial hierarchy system and humans in the species hierarchy system. Instead of seeing unity and a shared set of power among all races and all species, we continue to place white men at the top of our systems due to settler colonialism. We also continue to separate humans from nature.

Our Plant and Animal Relatives

This separation between humans and our environment—including both plants and animals—prevents us from seeing them as our relatives. This is why we continue to place economic values on them, as if we indeed saw them as our relatives, there would not be a price tag attached to them. It is no surprise that as Indigenous peoples we see our plant and animal species are relatives, hence, why many plants and animals play an important role in our creation stories. These creation stories differ among communities, tribes, and pueblos as we are distinct and not monolithic. For example, for us Zapotec people, our creation story states that our ancestors were born from trees and jaguars. Our ancestors were created by our deities inside caves by utilizing natural resources to create our people. We came from earth, and this explains the strong connection we have with our Mother Earth as Zapotec people. Given that we are children of deities, we believe that in our afterlife we return to the clouds and our spirits feed the earth with the most essential resource of them of all, water. This shows that we understand that our role on earth does not end when we are gone as we are continuing to provide water to our plant and animal relatives in the form of rain. This is why we call ourselves "cloud people" and this is the literal translation from our names, Binnizá, Binn Ditzaá, among other variations based on the Zapotec variant each pueblo speaks. This means that plants and animals are indeed our relatives as we came from them, and this is why we continue to protect them and advocate for their rights as well.

Our creation stories focus on native animals to our region, and this demonstrates that we always knew that we had to live harmoniously and in strong relations with them since time immemorial. Unlike Western European cultures that have commodified all natural resources, including animals, when we consume animals we continue to ask for their permission and protection before we consume them. However, these are not the beliefs or values Western cultures have as the agriculture systems focus on breeding animals for consumption in masses and making them undergo inhumane practices so that we can continue to have our beef and other meat products.

For Indigenous peoples, we have a relationship that Western cultures cannot understand with animals as they are our direct relatives, and we continue to understand that some of them are placed on earth for us to consume and only take what we need without overkilling or overhunting them. During colonization many of our animal species went through these *boom-bust* cycles because they were being overhunted and overharvested by settlers. This is the main reason why we witnessed huge declines of our traditional foods, in particular our animal species. We are still living in a time where we are trying to increase these species through Western conservation and its practices, but the continued impacts from climate change makes it hard. This is the main reason why we as Indigenous peoples know that Western cultures do not understand what it means to live in harmony with our plant and animal species as natural capital continues to value animals and plants for their economic value and not for the relationships one can build with them.

For my father, the banana tree became his relative, and this relative supported him while he endured the most awful times in his life. The banana tree had created his sanctuary away from the reality he was facing.

We strongly believe that the relationship he had built with this tree was why the tree protected my father. A bomb not igniting upon it being dropped is what Western religion teaches us is a miracle, but we deeply know that it was just nature taking care of my father because he had taken care of nature. If it were not for this tree, I would not have existed in this world and neither would this book. Many elders have told us that our ancestors once dreamed us into the future, but my father also believes that it is our plant and animal relatives that do this as well. So I will always wonder whether that banana tree dreamed me into my father's future and will be curious to know which of my descendants it also dreamed of.

"Kill the Indian in the Child"

My strong connection to my environment and the strong relationships with our plant and animal relatives my parents, grandparents, and relatives taught me to continue upholding pushed me to pursue the environmental sciences in academia. This is what inspired me to continue not only learning within the Western sciences framework but also to continue bringing to the forefront my communities and the knowledge they have passed down to me. This is the basis of this book as well, as I will bring both to the forefront. However, despite everything I have accomplished in academia, there are times when I thought I felt the *imposter syndrome*—feeling like I do not belong or have the intelligence to continue thriving in the environmental sciences. However, the imposter syndrome is justifiable for Indigenous scholars and scientists. Whenever I feel the imposter syndrome, I remind myself that this is okay because, at the end of the day, schools and academia were not created for us. It was created to assimilate our children (e.g., *boarding schools*), because there was one common goal during colonization and the introduction of Western settler states: to eradicate our Indigenous cultures and communities.

Boarding schools differed based on the country but were present throughout the Americas. In Mexico the boarding schools were created by heirs of those who created their wealth from the oppression against Indigenous peoples in the seventeenth century. The boarding school that was established and created in Mexico was *Casa del Estudiante Indigena* or Home of the Indigenous Student.[12] They recruited more than eight hundred Indigenous children for this school. This was considered an *experiment* because they wanted to find out whether teaching Indigenous children how to become models of nationalism would lead to its spread among Indigenous communities to assimilate them.[13] They taught them philosophies from Western European intellectuals because the founders and administration of this boarding school blamed all problems taking place in Mexico on Indigenous peoples. Many students went because education in rural communities, where most Indigenous peoples and communities resided, was not funded and nonexistent. However, most of the Indigenous students were forced to attend and were separated from their families. Like most boarding schools created to assimilate Indigenous children, they were mostly funded and administered by the Catholic religious regime.

When looking at the relationship my Indigenous communities have had with education, it has all been traced back to attempts of *assimilation* to eradicate our cultures and communities completely. However, they failed because Indigenous communities and pueblos continue to exist in modern times. However, knowing my history and experiencing the imposter syndrome throughout my academic journey have made me reflect on the purposes of education and how to continue maintaining my values as an Indigenous woman. Now that I reflect on those instances when I felt the imposter syndrome, I can point to the racism that was enacted against me.

I started having these reflections as I got older because when I was younger, my perspective was different. As a displaced Indigenous woman, I lived my younger years trying to seek validation under the Western notions that govern our current native lands. I was fed this idea that the only way to do this was through higher academia. Being raised in the diaspora, in a low socioeconomic household, and in South Central Los Angeles, I was taught in school that my education was the only way out of poverty. However, the more I immersed myself into academia and higher education, the more I realized that my elders, relatives, and parents hold more knowledge about our environment than any professor, scientist, or researcher I have come across in my journey. We do not need validation from settler colonialism because we have been living in harmony with our environments.

As mentioned before, my grandmother's teaching to live in harmony with not just the land but also the Indigenous peoples and communities on whose land I currently reside to become a welcomed guest is something I continue to sustain in the diaspora. It is why despite moving cities to attend graduate school or for employment reasons, I always end up working with Indigenous communities. When I was employed in Louisiana to work on the Deepwater Horizon oil spill, I ended up building relationships with the local tribe, the United Houma Nation. In Seattle, I have worked with the Duwamish people and urban Indigenous peoples who now reside there. Building these relationships with Indigenous peoples whose land I reside on as an unwelcome guest continues to allow me to not only build this harmony but maintain an important cultural value I uphold as a Zapotec and Maya Cho'rti' woman.

Invalidating Our Indigeneity

Ironically, our firsthand experiences as Indigenous peoples are invalidated by professors, scientists, and researchers who maintain power and privilege in these educational systems. While exploration is what drives Western science, the exploration and questioning of our environment has already been done by our ancestors. That knowledge our ancestors formulated has been passed down through oral stories, songs, prayers, and other traditions. Thus, we hold that powerful Indigenous knowledge or Indigenous science as I refer to it. This is the same Indigenous science that can heal our Indigenous lands because we have been formulating our knowledge systems longer than settlers have on our lands. My academic journey has been bestowed within the environmental sciences.

An example I want to share that demonstrates how our Indigenous science is different is how we view invasive species. The Western sciences teach us that invasive plants are pests, unwanted, or do not belong in this landscape. However, to us, invasive plants are displaced like many of us. They were forced from their native lands and like many of us had to adapt to a new environment. Like banana trees, we are forced to uproot ourselves from our native lands and have to adapt to new environments. Some invasive species have thrived so well that many folks think they are native species because of the abundance of them in many landscapes. While not forgetting the negative impacts they do have on some environments, some of our Indigenous communities have learned how to live with them.

For example, we Oaxacan and Salvadoran folks use a lot of banana leaves to make tamales and other traditional foods. While bananas are not native to our homelands, they have become our relatives that continue to nourish us in various forms. Like banana trees, for those of us who are displaced from our homelands, we build on a strong foundation, our roots, and establish our new ecosystems and communities in the lands we now reside on. Our resilience is adapted to our current environment, and banana leaves serve as the metaphor to our descendants. We are both strong and resilient despite being displaced and have become an essential component of Indigeneity within the US context.

Unfortunately many of our Indigenous teachings, like the one my father has passed down to me, are not respected by Western science. Our teachings are often considered invalid because they do not fit or follow the linear way of thinking that our Western scientific method follows. However, as a trained environmental scientist, I believe my Indigenous teachings hold more weight in my life than my bachelor's or master's and doctoral degrees do. My Western credentials are not the reason why I am alive, why my family's legacy continues. Protecting and respecting nature and honoring our kinships with Mother Earth are the reason I can stand here today as an Indigenous woman and scientist. I write this book to provide an analysis and personal narrative of my journey as an Indigenous woman and scientist who owes her life to nature. Culturally grounding my work allows me to understand that in order to heal our Indigenous landscapes, we have to do it through our Indigenous science.

Longing to Return Home

There is a longing we continue to feel as displaced Indigenous peoples to return home to our ancestral native lands. This longing never leaves us as being displaced is something we are reminded of every day. However, our ancestral homelands and environments continue to carry the trauma that pushed many of our grandparents and parents out. It is hard reading headlines of El Salvador being classified as the world's most dangerous country or a country overrun with gang violence. I think that there is a lot of judgment by outsiders of the trauma and how our people coped with this trauma, particularly from the United States. However, these narratives fail to acknowledge the role the United States played within the war that lurked over El Salvador. While we build harmonious relationships with our environments in the diaspora, nothing is the same as being home as those relationships are ingrained in us since birth. The war trauma manifested differently to my dad as it did not push him to become violent, but this just shows that war trauma impacts individuals differently. When I asked my father how war impacted him, this is how he responded:

> Not very well. War is never something light to take. It mentally impacts us in a negative way. I healed over the years from war but we still hold on to that trauma. War leaves us thinking that we will never be safe, even now while being displaced. I still do not feel safe. War instills this fear we carry on for years. It has taken me years to come in peace with war and the impacts it left on me. I do not recommend war to anyone, it is bad.

For my dad war manifested a fear that he still carries today. However, for many, war instills a longing to be violent or show violence when one faces a barrier or any threat. Violence becomes a protection mechanism that some of our grandparents or parents taught us. While this was not the case for me, I understand why many of our Salvadoran youth created gangs in the diaspora. The *Mara Salvatrucha,* or MS-13, is a gang that originated in Los Angeles in the 1980s. Many of its founders were immigrants that had come escaping the war from El Salvador as it was still ongoing during these times. Many of our relatives came in contact with the current gangs that existed in Los Angeles during these times. These gangs were harming or threatening many of them and their families. Thus they decided to form a gang, similar to how the guerrilla was formed to create groups to protect themselves and one another.

For many of them, violence is how war impacted them and this was the only way they knew how to protect themselves. The United States decided to deport many of them if they were caught with any gang activity, and as a result, their violent tactics have spread in El Salvador. Their violence or murders are not justifiable, but their existence shows how war impacts people differently. Knowing that our country of El Salvador is still overrun with violence, the aftermath of the civil war makes it hard to cope with the longing my father feels as safety is something he prioritizes now. Every time he returned to El Salvador, the life he endured during the war resurfaced. This is why I know our environments carry our trauma as they hold those memories for us. He has to cope with this longing to return and his longing to feel safe because, like he mentions, he still does not feel safe today. This is why it is up to my generation to start our healing journey as war trauma lurks within us every day. Like banana trees, we must grow new banana leaves and continue to build a new chapter in our lives to ensure we can heal as displaced Indigenous peoples who had to also endure a war, a genocide against our people.

2

Ecocolonialism of
Indigenous Landscapes

Our relationships as Indigenous peoples with our land have been altered as a result of settler colonialism. Settler colonialism introduced so many layers onto our societies that continue to favor *whiteness* and *white supremacy*.[1] Whiteness grants those who are white the power and privilege due to the racial hierarchies settler colonialism introduced. It is the system that continues to determine that white individuals dominate our political structures and leadership. In the United States whiteness is what made many European immigrants, those who came after the thirteen colonies were formed, assimilate to the white American persona. Whiteness embodies the white American persona and identity that continue to be associated with the United States, even globally. This is why as Indigenous peoples, when we travel abroad and tell people that we are *American* they question how this can be if we are not white. This is the whiteness that continues to dominate what an *American* should look like.

On the other hand, white supremacy is a set of toxic systems that ultimately favor white men and women and grant them the power to enact any form of violence or oppression in order to maintain whiteness in the Americas.[2] White supremacy upholds whiteness's superiority, and this is the set of ideologies that settler colonialism continues to uphold within our Indigenous landscapes. This means that if you are white, you have more power over managing our natural resources, and this is why it is not a surprise to Indigenous peoples that environmental policy makers, scientists, organization leaders, and others who have a seat at the table and lead this table are white—mostly white cisgender men. When white men and people who have political power serve as governing agents over our natural resources and do not respect Indigenous sovereignty, they are practicing a form of *ecocolonialism.* I refer to ecocolonialism as the three layers from settler colonialism that continue to impact our environment and landscapes. These three layers include

1. White people governing over our natural resources and Indigenous lands without consulting Indigenous peoples of those lands or respecting Indigenous sovereignty;

2. The severe altering of our landscapes due to settler colonialism and the ideologies it introduced, including climate change impacts; and

3. The lack of resources offered to Indigenous or communities of color who are already experiencing the impacts of climate change that oftentimes results in displacement.

While I break down ecocolonialism into three layers, there is a lot to dissect beneath these layers as they embody a myriad of impacts and oppression. Ecocolonialism ultimately is the altering of our environments and landscapes due to colonization of Indigenous lands and the paradigms that are upheld to grant settlers (white people) the power to continue managing our environments and landscapes. Every environmental scientist, researcher, policy maker, and anyone else who ever took a decision over our environments without consulting the Indigenous peoples of those environments has practiced some form of ecocolonization.

Ultimately, these hierarchies and power structures embedded in our environments continue to displace Indigenous peoples from their ancestral lands as they continue to face oppression and violence when their environments are severely altered and impacted, because *place* is crucial to our Indigeneities.

Our place or environments are what uphold our cultural values that ultimately make up our identities as Indigenous peoples. For me personally, both Oaxaca and El Salvador are important places to my Indigeneity because they are the landscapes that manifest through the dualities of my identity. As a Zapotec woman from Oaxaca, my pueblo and ancestral environment manifests through my teachings, language, and culture. Also, as a Maya Ch'orti' woman from El Salvador, my ancestral land and teachings from my environment manifest through the other teachings, language, and culture that I embody as a transnational Indigenous woman. Indigeneity is a complex identity, and as someone who has a duality within her Indigeneity, my identity becomes more complex. The importance of our native lands and place is why displacement is something important to mention if this is something we have undergone as Indigenous peoples. The displacement of Indigenous peoples from their ancestral homelands is also another way ecocolonialism manifests. However, our Indigeneity is still maintained, even in the diaspora. Personally, I believe some assets of our Indigeneity adapt because we have to in order to survive and thrive in our new environments.

Displacement beyond Borders

Displacement and diasporic narratives mostly focus on those who have been displaced among borders. These borders were created by colonizers to ensure their nationhood as they *conquered* Indigenous lands throughout the globe. In North America, these borders are what separate Canada from the United States and the United States from Mexico. However, it is also important to bring to the forefront the diasporic narrative that is often ignored—the narratives of Indigenous peoples who are displaced from their rural spaces, mostly reservations, into urban spaces, cities.

Most tribal reservations that were established in the United States are located in rural spaces and environments. Sometimes they are hundreds of miles away from the closest city. Rural places have a low population density and more land than infrastructures. These rural spaces also continue to face a lack of resources or aid provided by the state as the populations are smaller and, when it comes to reservations, entirely embodied by Indigenous people. These two indicators, population and race demographics, are the determining factors that dictate why most funds are allocated to cities instead of rural places.

While I have discussed our displacement through physical colonial borders, I also want to address our displacement from rural environments into urban environments that are not necessarily defined by colonial borders. This displacement is important to highlight because despite Indigenous peoples moving away from their reservations, whether it be through borders or nonborders, displacement still impacts us. It may fracture our *Indigeneity,* Indigenous identity, but it never vanishes. Indigeneity is carried throughout ourselves, despite being displaced from our ancestral lands. Focusing on displacement beyond borders foregrounds the importance of urban Indigenous people, or oftentimes referred to as *urban Natives.*

Many of us are forced to relocate to urban spaces, cities, in order to access better educational opportunities. In Oaxaca, many of our Zapotec people move from their rural and small pueblos to larger cities in Mexico in search of better jobs or educational opportunities. While colonial borders separate Indigenous peoples of Mexico from Indigenous peoples of the United States, this displacement that tends to happen more often from rural places to urban places is very common among both narratives. During my doctoral program at the University of Washington, I was working on a restoration project in an urban park. Given that this park was an important space for urban Natives, I wanted to explore the stories behind them to understand the diaspora from within the country, not necessarily across borders. The following response demonstrates the common theme that most responses incorporated.

> **Why do you live in the city?**
>
> *Diana:* So I moved here three years ago to go to school, which ties back to my idea of access to opportunities. I knew that I wanted to leave the reservation to go to school. I'm very glad that I'm here.... I really like Seattle.... There are Indigenous peoples from all over, different places.

Many urban Natives who I interviewed relocated to the city, in this case Seattle, to have better educational and job opportunities. It is no surprise given that most universities, larger corporations, and other entities are located in cities. It is also important that while it may be rationalized as a choice given to Indigenous peoples to have the option to move to cities, it is often more of a forced choice. It reminds me of how my father phrased his decision to join the war as a *forced decision* rather than an option, because, oftentimes, Indigenous peoples are faced with ultimatums. They either continue to endure their current and existing harsh conditions or they choose the option that grants them some security from these same harsh conditions. My father either had to choose to either continue to be persecuted by the *escuadrones de la muerte* or join the guerilla to gain some form of security and safety. This parallels with why many Indigenous peoples have to leave their rural places of origin to relocate to cities. The lack of job, educational, and other essential opportunities forces them to make a choice to move to a city.

These same options were enacted into laws by the United States during the 1950s. For example, the Indian Relocation Act of 1956 encouraged Indigenous peoples to relocate to cities so that they could take advantage of employment opportunities that were nonexistent in their tribes.[3] However, despite whether this relocation or displacement is across colonial borders or within the same confinements of these borders, relocation and displacement have an impact on Indigenous peoples. Dr. Daniel Wildcat identifies the *three removals* in his book *Red Alert! Saving the Planet with Indigenous Knowledge* as being *geographic, social,* and *psycho-cultural.*[4] These removals were displacement or relocations Indigenous peoples had to endure, and while he contextualizes this narrative within the United States, it is applicable in Latin America as well.

Many Indigenous peoples of our pueblo have to take on this *internal relocation* that still impacts them, despite it not being an international relocation. Thus, oftentimes, a generation also is not given a choice as a previous generation's relocations determined that for them. If one's parents or grandparents relocated to the city, it is not rare that the city becomes all we know as urban Indigenous peoples. It is important to mention that internal displacement can occur because one chose to relocate or that choice was already made for us by the previous generation. As part of the interviews I conducted with urban Natives residing in Seattle, a previous generation's relocation was something that came up constantly in many testimonies.

Previous Generation's Relocation

Why do you live in the city?

Selena: I grew up in Seattle.... I really go out of my way especially when it comes to the urban Native community and sharing the fact that especially with the urban Native sphere, there is a lot of solidarity work, a lot of community, and when I'm telling myself how to carry myself within these spaces, I always revisit the concept that I serve as a community liaison, for that 1 percent of the US population that identifies as Indigenous.

According to the United States Census Bureau, over 70 percent of Indigenous peoples live in urban settings. This is something important to keep mind when discussing Indigenous peoples' relationships with land, water, and environments. However, one thing that is important to mention is that despite the relocation from rural to urban settings, our Indigeneity is not lost. Our Indigeneity adapts, something settler colonialism still fails to understand. We adapt to our environments and to contemporary times, the same way we can adapt to climate change impacts that are altering our environments and landscapes. Colonialism aimed to make our cultures disappear, which is why photographs of our people were taken, our artifacts were stolen to preserve in museums, and our stories or way of life was documented by anthropologists. However, we are still here thriving despite the colonial legacies that continue to govern this country. Whether we live in an urban or rural place, we are still holding the forefront of what it means to be Indigenous in the twenty-first century.

> *George:* I feel like it's very important for me to remember that Seattle is a rainforest, so acknowledging that we're living in an area that has been developed but still has a lot of environment and nature right there in the forefront versus looking at it as being a part of the background.

Cities Are Indigenous Lands

Your homelands can also be what are now known as cities. For example, Seattle is located on Duwamish lands. It is the ancestral lands of the Duwamish people.

This continued dialogue of urban versus rural places also fails to recognize urban places as Indigenous lands. An observation I made while conducting these interviews for my dissertation is that even as Indigenous peoples, we tend to think of urban places as non-Indigenous locations given the extreme changes these landscapes have undergone as a result of urbanization. The altering of our environments that has completely changed them is what I refer to as ecocolonialism. Ecocolonialism is the history that weaves together the ecological changes that took place on our Indigenous lands starting since colonization. Ecocolonialism explains how settler colonialism has impacted and continues to impact our landscapes and environments. This, coupled with climate change, continues to jeopardize not just our environments but also our way of life.

Many cities like Seattle are built on Indigenous ancestral lands. Despite being severely altered and completely changed in some instances, they continue to be the ancestral lands where many burial sites of our ancestors are. Seattle, for instance, is built on Duwamish lands, so Seattle has an interesting history with the Duwamish people. It was named after one of their chiefs, Chief Sealth, who, in 1865, was not allowed to step foot in this city after ordinance law number 5 was passed.[5] This law banned Native Americans from the city limits. It was a way to get rid of the Indigenous peoples of these lands in order to make space for what the colonizers had in mind to do with this city. The city that was named after a chief also became the same city that did not permit him or his people to step foot in it. The irony is that many people who reside in Seattle do not know this history because, unlike settler history, Indigenous history is something folks have to research. It is not as accessible or even integrated within the educational system the same way colonial history is.

Seattle, known as the Emerald City because of its greenery scenery, is named after an Indigenous man who died without being able to ever step foot in the city again. His people to this day have also not received tribal federal recognition. Many cities continue to bury Indigenous history because acknowledging it will bring out many injustices that have been taken against the Indigenous peoples whose lands cities are built on. However, this history needs to be acknowledged in order to address the hurt and harm that has been caused against Indigenous peoples since colonialism.

While cities can provide its residents with many accommodations that are not available in rural places, Seattle serves as an example of how cities also carry a deep history that continues to erase Indigenous peoples from these locations—both the peoples whose lands cities are built on and the current Indigenous peoples residing in these cities. When I reflect on the city life and all the amenities it offers, I cannot help but reflect on the landscapes that the Indigenous peoples of these lands stewarded and cared for since time immemorial. Ecocolonialism has completely destroyed these landscapes that once were flourishing and are now gone because the city has taken over the entire landscape. The construction of these metropolitan areas required our landscapes to be severely altered and heavily destroyed. Looking back at the impacts of both displacement from our ancestral lands and the ecocolonialism that has destroyed and altered many of our lands, all I can ask now is, How have they impacted our relationships with our environments and nature? In order to answer this question, I need to go back to the three layers of ecocolonialism to define some ways it has impacted us as Indigenous peoples:

1. White people governing over our natural resources and Indigenous lands without consulting Indigenous peoples of those
   lands or respecting Indigenous sovereignty

In Mexico, as I mentioned in the introduction, whiteness in this country favors white Mexicans and mestizos. If we look at the racial caste system, we can see how the *peninsulares,* Spanish aristocrats, have the most power and privilege.[6] They are followed by the *criollos,* Spanish aristocrats' descendants. Following them are *mestizos,* Spanish and Indigenous descendants, and *mulattoes,* Spanish and Black descendants. On the bottom of this racial caste are *Indigenous* people and *Black* people. I break down this racial caste system into the six power structures, starting with the highest power and privilege: (1) peninsulares, (2) criollos, (3) mestizos, (4) mulattoes, (5) Indigenous, and (6) Black.[7] It is also important to note that these racial terminologies are offensive as they were created as a result of settler colonialism and should not be used, especially when referring to Indigenous and Black people, who continue to be racialized in Mexico for being Black and/or Indigenous.

However, Indigenous and Black people who *mixed* with any Spanish blood gained more access in the racial caste system Mexico established as their main goal was to *water down* or *dilute* Black and Indigenous blood. It was the blood quantum narrative Mexico portrayed for so many years and continues to uphold in its current systems of power.[8] In Mexico, many people can quantify their Spanish blood as being a descendant of a Spanish aristocrat, which is still highly esteemed in the society. Yes, it is more common that Indigenous and Black Mexicans have some "Spanish blood" but despite this, the racial caste system continues to not favor us. The only Indigenous president Mexico has ever had was Benito Pablo Juárez García. He was the twenty-sixth president of Mexico and held office from 1858 until 1872. He was born in Sierra Juárez to two parents who were Zapotec peasants and farmers. Being the first and only Indigenous president thus far makes him an important political figure that Indigenous peoples continue to honor to this day.

The other successful presidency run we had that supported an Indigenous presidential candidate was in the 2018 general election where María de Jesús Patricio Martínez, more commonly known as Marichuy, ran for president. She was the candidate supported by the National Indigenous Congress (CNI), the Indigenous caucus and council we have established in Mexico, and is a Nahua Indigenous healer and medicine woman.[9] Unfortunately, she lost to Andrés Manuel López Obrador, a *mestizo*.[10]

While Benito Juárez broke some of these racial barriers that continue to exist in Mexico that favor Spanish aristocrats, their descendants, and mestizos, there has not been another Indigenous president since 1872, over 149 years ago.[11] The same goes for all political powers and agencies in Mexico, so white people and, in Mexico, *mestizos* continue to govern our natural resources and dictate how our land can be used. The same notions occur in the United States, which has a more binary racial caste system that is often denoted in the US Census. Whites continue to have political power to make decisions for natural resources without having to consult Indigenous peoples. The only difference in this power hierarchy is that tribes and many Indigenous communities in the United States have a unique positionality due to *tribal sovereignty*. According to the National Conference of States legislatures, tribal sovereignty is defined as "the right of American Indians and Alaska Natives to govern themselves. The US Constitution recognizes Indian tribes as distinct governments and they have, with a few exceptions, the same powers as federal and state governments to regulate their internal affairs. Sovereignty for tribes includes the right to establish their own form of government, determine membership requirements, enact legislation and establish law enforcement and court systems."[12]

This grants tribes, nations, and pueblos in the United States their autonomy to make political decisions over their sovereign lands. However, this only applies to federally recognized tribes, and many tribes like the Duwamish tribe, the original stewards of what is now known as Seattle, do not have federal recognition. This limits their right to govern themselves and receive protections, services, and some benefits that the United States has agreed to provide federally recognized tribes. The US Department of the Interior Indian Affairs has determined that in the United States there are currently 574 federally recognized tribes.[13] However, tribal sovereignty is oftentimes not respected or taken into consideration by governing agencies that have a strong political presence when it comes to making decisions about our environment. In the state of Washington, one of the cases I witnessed that ignored tribal sovereignty initially was the ballot initiative I-1631.

When we look at urban settings and built environments, we tend to forget that these are also built on Indigenous lands. These Indigenous lands continue to be the ancestral homes to current Indigenous communities that are now positioned at this unique intersection. They are living on their ancestral lands that have been completely changed to create metropolitan areas such as Seattle, Los Angeles, and New York City. For Indigenous communities who find themselves at this intersection, they are more often ignored and forgotten than communities who live in more rural areas. Organizations, government structures, and other entities tend to forget to even consult or include these communities into their legislative work. Too often many organizations and other entities mention that due to time-sensitive documents or projects, they were not able to consult tribes from rural areas, forgetting to acknowledge that Indigenous peoples and communities also reside in cities. For example, in Seattle we had ballot initiative I-1631, which was spearheaded by the Nature Conservancy in 2018.[14] Initially this ballot initiative did not include Indigenous communities, specifically the tribes that have tribal sovereignty and operate under this framework that allows them to have some self-determination over what takes place in their tribal nations and reservations. The Nature Conservancy explained the bill this way:

> I-1631 is a practical first step to ensure clean air and clean water for everyone in Washington and gives us the chance to pass on a healthier state to the next generation. It will create good jobs and invest in clean energy like wind and solar, healthy forests, and clean water with a fee on pollution paid by the state's biggest polluters.[15]

However, initially tribes were not consulted, including the Duwamish tribe, who continues to reside in the city of Seattle. It was not until tribal members started calling on the Nature Conservancy to include and respect their tribal sovereignty that they started to include Indigenous voices. Situations like this continue to demonstrate that Indigenous communities and people continue to be afterthoughts in things like politics, research, and projects. Healing our Indigenous landscapes can begin with simply starting to center Indigenous peoples and placing them on the front of everything as opposed to continuing to treat them like footnotes and afterthoughts.

Climate change and environmental policies continue to take place in cities; however, urban Indigenous peoples are not included or even consulted. I strongly advocate that in order to heal our environments, we must shift our focus to curb climate change from within cities. However, in these policies Indigenous peoples must be placed front and center as should other communities of color.[16] Most importantly, tribal sovereignty needs to come to the forefront of these initiatives and policies; otherwise, it is a form of ecocolonialism that continues to grant only white people the right to govern over our lands.

Indigenous Pueblos of Mexico's Tribal Sovereignty

The basis for tribal sovereignty in the United States is mostly due to the treaties that the Indigenous tribes and nations signed with the United States government. However, in Mexico, Indigenous pueblos and communities never signed treaties with the Mexican government. Therefore, there is no set tribal sovereignty in Mexico to legally protect Indigenous pueblos and communities. The first conversation around Indigenous rights that took place within the Mexican government was in 2001 when it decided to ratify its constitution, in particular articles 1, 2, 4, 18, and 115 to grant rights to Indigenous peoples.[17] These articles were amended to include the right to self-determination and autonomy to Indigenous pueblos. It also incorporated Indigenous rights that each Mexican state must respect in particular when it comes to getting consent and permission from Indigenous communities before major decisions are made to their land. It highlighted their rights to also govern their territories, languages, culture, religion, and other elements that are important to their pueblo's identity.[18]

While the Mexican government tried to define who was considered a *pueblo originario,* Indigenous pueblo, there has been no official definition because each pueblo is distinct and not monolithic. The Mexican government incorporated the definition in the Mexican constitution that defines Indigenous pueblos as the original communities and societies that have survived precolonization and were impacted and continued to be impacted by colonization. Conversations around Indigenous rights and sovereignty were further advanced when the United Nations (UN) introduced the UN Declaration on the Rights of Indigenous Peoples (UNDRIP) in 2007. However, despite these amendments to the Mexican constitution, Centro de Estudios de Derecho e Investigaciones Parlamentarias (CEDIP) released a report in 2017 that states that Indigenous people continue to face violation of their land rights that result in violation of their rights, threats, hostile situations, and even the murder of many Indigenous leaders who speak up.

Article 2 of the Political Constitution of the United Mexican States (CPEUM) raised Indigenous rights to a constitutional status that was initially recognized through Convention 169 of the International Labor Organization (ILO).[19] The amendments to this article reiterated the recognition of Indigenous peoples and communities and granted them the right to self-determination and autonomy. The amendments made to articles 1, 2, 4, 18, and 115 also include the recognition of Indigenous peoples. It is important to note that Mexico had not recognized Indigenous peoples because of its notions of eradicating Indigenous peoples in this country through *mestizaje.* During times of colonization, the Spanish crown separated Indigenous peoples from the new Hispanic population to avoid this notion of *mixing* within races. However, the mixing of races was inevitable and gave rise to the racial caste system, which was made up largely of people who came from "illegitimate interracial unions" of Spaniards, Indigenous peoples, and Black people.[20] Interracial unions were not accepted in Mexico, so many people were racialized as mestizos. This goes back to the new racial caste developed postcolonization, previously mentioned, that places power and privilege according to Spanish blood. These include (1) peninsulares, (2) criollos, (3) mestizos, (4) mulattoes, (5) Indigenous, and (6) Black.

Mestizaje continues to be used as a narrative to claim that in Mexico, everyone is the same and that, if anything, the Mexican society is only classist because mestizaje portrays the Mexican identity in the twenty-first century as mixed heritage that embodies whiteness (from Spain), Indigeneity, and Blackness. However, Indigenous and Black people know that this is the reality as racism exists in Mexico. The racial caste continues to favor peninsulares, criollos, and mestizos over Indigenous and Black people. However, it is important to mention that there is racism in Mexico against Indigenous and Black people, and despite these amendments made to the constitution, there are years of advocacy and leadership that Indigenous communities led to get this accomplished.

One of the Indigenous-led groups that continues to lead these advocacies is the *Congreso Nacional Indígena* (CNI), translated to National Indigenous Congress.[21] The National Indigenous Congress, founded on October 12, 1996, was formed because Indigenous pueblos of Mexico came together wishing to create a safe space where they could reflect and build their solidarity among other pueblos to strengthen their resistance. They wanted to separate themselves from the government and have their own forms of organization, representation, and decision-making. The Indigenous pueblos and people that make up CNI include Amuzgo, Binnizá, Chichimeca, Chinanteco, Chol, Chontal de Oaxaca, Chontal de Tabasco, Coca, Comcac, Cuicateco, Cucapá, Guarijío, Ikoots, Kumiai, Lacandón, Mam, Matlazinca, Maya, Mayo, Mazahua, Mazateco, Mixe, Mixteco, Nahua, Ñahñu/Ñajtho/Ñuhu, Náyeri, Popoluca, Purépecha, Rarámuri, Sayulteco, Tepehua, Tepehuano, Tlapaneco, Tohono Oódham, Tojolabal, Totonaco, Triqui, Tzeltal, Tzotzil, Wixárika, Yaqui, Zoque, Afromestizo, and Mestizo.[22]

CNI was formed after the rise of the Zapatista Army of National Liberation (EZLN) that we will learn more about in chapter 6.[23] One would think there was probably not a high need for this council because the constitution grants Indigenous peoples their rights to govern themselves or be consulted over their land. However, most of these amendments to the articles of the Mexican constitution are not respected or validated among many Mexican states. While some states intend to ensure they follow it, Indigenous pueblos and people of Mexico constantly have to fight for their rights. This need to fight for their rights also extends to Afro-Mexicans and Afro-Indigenous peoples.

Fighting for Indigenous Land Rights

Oftentimes, as Centro de Estudios de Derecho e Investigaciones Parlamentarias (CEDIP) has reported, Indigenous people continue to face violations to their land rights that result in threats, hostile situations, and even assassinations because most of the people who violate their land rights have political power and money to buy their way through the corrupt governments of Latin America. They go against large agricultural corporations such as Monsanto, which was one of the largest multinational agricultural business companies in the United States. It was founded in 1901 and primarily engaged in the production of herbicides and genetic engineered seeds.[24] Given that the United States has more strict land regulations than other countries, like Mexico, they often did business abroad. In my lens, they were one of the driving agribusinesses that continued to practice ecocolonialism because their herbicides and genetically modified seeds drastically impact entire ecosystems.

This is what the Mayan Ich Eq community was facing in Mexico. Leydy Pech is a Maya Ich Eq woman that led her community, located in Hopelchen, Mexico, into opposing Monsanto.[25] Monsanto, like other large agricultural businesses, had been purchasing lands along the Mexican coast, thus displacing Indigenous peoples further from their lands. Not only did Monsanto and these large corporations clear the rainforests that once were home to many species in these lands, but they also started using chemicals and pesticides to grow soy.[26] The habitat loss, coupled with the pesticide usage, was severely impacting the endangered bee species native to the Yucatán, *Melipona beecheli*. This bee is an integral component of their communities because not only is it a relative to the people but it is also an integral component of their economy.

Many people in Leydy's community are beekeepers who live off the honey of these bees under their care. The relationship they have with the bees is unique and very different from what Western society has developed in terms of beekeeping. This goes back to my father's teaching that nature takes care of us if we take care of nature. Bees, *xuna'an kab*, do not sting them because they are their caretakers.[27] They feed the bees and build strong relationships with them where they are not exploiting the bees just for their honey. They are referred to as *xuna'an kab* or *colel-kab*, which translates to royalty or queen. This demonstrates how much they are praised, as in our Indigenous languages, the names we have for animals sometimes embody the relationship we have with them. If the bee is considered royal, this means that they are highly esteemed by the Mayan Ich Eq community, so it is not a surprise that they symbolize fertility.

They are also endangered, and scientific reports from 2005 claimed that this bee species could become extinct.[28] This is why the community kept working hard to protect it, as it is not only an integral part of the ecosystem but also their cultures. Their care and love for the bees are why they followed Leydy and advocated for the bees' well-being and safety. The beekeepers maintained a strong relationship with the bees as they still say that they can feel and experience the same things the bees do. For instance, not only were the bees being affected by the air fumigation but the beekeepers were getting sick as well. This is why it was no surprise that since the bees were sick and dying, so were the people.

In Mexico, the government needs to consult the Indigenous peoples when selling their lands to large corporations and introducing genetically modified crops, as the articles that were amended in 2001 state. While Leydy and her community were discredited by the Mexican government and faced harassment and doxxing, they continued fighting for their rights and, most importantly, their bees. They were able to make changes to protect their bees, but this was not easy and required years of litigation for them to be heard and acknowledged. Leydy was an integral part of this movement, and it is amazing that she is finally being recognized for her work as an Indigenous woman who saved her bees and community. Her community's story is a reminder that despite the Mexican constitution recognizing Indigenous peoples and their rights to self-autonomy, they continue to be ignored and not consulted when it comes to their land and territories. This is why many Indigenous pueblos of Mexico mention that the constitution intended to make a difference for their communities but has not done so in actually applying these changes into actions since 2001.

> *The constitution tries to make an attempt to respect*
> *Indigenous self-determination and our autonomy*
> *but it is not really respected or validated.*
>
> —LITA SER, founder of Ñaa Ñanga Tijaltepec

2. The severe altering of our landscapes due to settler colonialism and the ideologies it introduced, including climate change impacts

Ecocolonialism has introduced a different type of loss that we are continuing to experience as Indigenous peoples. This loss extends beyond just human loss to animal, plant, and other environmental loss. All environmental loss Indigenous peoples experience results in cultural loss that fractures our identities. Like all loss, it results in grief, and the loss that results from ecocolonialism and climate change is what I refer to as *ecological grief*. Deforestation results in the loss of plant and animal relatives and this is why when landscapes experience this clearing of trees and biodiversity to make space for large agricultural companies, this is another way ecological grief manifests. When animals get sick, like the *Melipona beecheli* (bee) did, the community also experiences ecological grief.

To explain how ecological grief feels to many of us, I ask you to think of a pet you loved. When this pet passed away, you felt sadness and grieved their passing. This grief is similar yet distinct from the grief we experience when we lose our relatives because we are grieving for an animal. For Indigenous peoples, we have these similar feelings but not just for our pets; we also have it for other animals and plants that are a part of our environments. This ecological grief is rooted in the relationships we hold with nature and our environments and how we manage these resources when they are under our complete stewardship and care. To further explain ecological grief, I will share the story of the Mixtec, *Ñuu Savi*, community in San Pablo Tijaltepec, Oaxaca.

In San Pablo Tijaltepec, Oaxaca, heavy rains destroyed the communal harvest, grown in their *milpas* or *ituu*, in fall 2020. Toward the end of 2020, Oaxaca was still recovering from the earthquake that took place the summer of the same year—they are still recovering from the earthquakes that took place in 2017 to this day—and the impact of the 2020 earthquake forced many families to start again from ground zero. If this was not enough to increase Indigenous peoples' vulnerabilities in the Oaxacan region, the heavy rains they experienced in 2020 ended up flooding rivers as well, thus flooding many homes.

However, one of the biggest impacts the rains had was on their food sources and crops such as their corn, beans, and squash. As mentioned earlier in the book, these crops are primarily grown in *milpas*. Milpas are sustainable systems that are corn fields, but unlike the corn fields in the United States, milpas are generative small agricultural systems that date to our ancestors—an ancient traditional way of growing crops that our Indigenous communities have sustained, way before the colonizers introduced their extractive agricultural processes. Milpas also create small ecosystems for other plants such as beans, squash, and small animals our Indigenous pueblos consume. These plants and animals have positive interactions with corn, so it builds a reciprocal system that benefits all plants. Some of the small animals include armadillos and crickets, which are a source of protein among many Indigenous Oaxacan traditional diets.

This agricultural system is why Indigenous communities of Mexico, primarily Oaxaca, have been able to sustain and maintain the existence of over fifty-nine different types of corn. The primary crop grown in milpas is corn, and corn is what allows its relatives, such as peppers, beans, and squash, to grow beside it. Milpas are sacred and they mean life to many

Indigenous communities of Oaxaca. Every harvest, we are taught to ask for permission from our god to cultivate what the milpa has offered us.

During the months of March and April, the milpa ends up burning itself because of the extreme high temperatures. Before this happens, we go to the milpas and start asking all the animals that are living within it to please leave because the fire is coming and it will end up killing them. If we find any animal that has been killed in the fires, we offer something to the milpa in return for the spirit it took with that animal, and we also say a prayer thanking the animal for sacrificing itself. We know that the animal is going to decompose and provide nutrients to our milpa. Once the rain season returns, we do our rituals and ceremonies to our gods or goddesses and ask them to replenish our milpas with crops that will nourish our mind, spirit, and body. In August and September we begin planting our maize (corn) again.

The milpa becomes a communal harvest that our entire pueblo takes part in.[29] As a result, it also represents the heart of our pueblo as we all play a role and dedicate our time and energy to ensuring that the milpa grows. The milpa is a diverse and complex ecosystem that provides us with many of our traditional foods. This is where most Oaxacan people collect their grasshoppers, which is an insect that plays a vital role in our traditional diets. The milpa also offers us other animals and foods like armadillos and different types of mushrooms that we integrate into our diets. Every living thing in the milpa is our relative, and they all coexist with one another. This coexistence is also beneficial and mutual as they support one another with nutrients, adequate shade, and other things they might need.

Due to their complexity, milpas are able to protect themselves from plague. However, in some instances, we have witnessed how they have not been able to protect themselves from plagues or diseases that climate change brings to our communities. Our pueblos try their best to avoid pesticides as this contaminates all the food sources the milpas offer us, ultimately, harming us as well. As long as we have access to growing our own milpas and maintain this knowledge, we will continue to have access to our traditional foods. This is why milpas have become a crucial component of our food sovereignty and justice in Oaxaca. However, one thing our milpas are not able to protect themselves from is heavy rains and other natural disasters that end up hurting them.

This is what happened to many milpas in different communities across Oaxaca. When the heavy rains destroyed their milpas, the first thing that came to their mind was not the fact that they had lost their communal harvest or foods. Many pueblos were extremely sad and mourned for their milpas, grieving the loss of their plant and animal relatives in week-long rituals and ceremonies as several spirits were lost. It was a sadness that clouded the entire community.

Unfortunately, with climate change, we are experiencing more ecological grievances because, as Indigenous peoples, we experience firsthand the environmental degradation and impacts climate change has on our plants, animals, and other nonliving things that are an integral part of our ecosystems. Now imagine not just losing one pet, but multiple pets in a couple of hours. Ecological grief at this magnitude is why many of our Indigenous communities continue to be on the frontlines of climate change. I think many still fail to understand this. A lot of environmental scholarship that mentions Indigenous peoples as frontline communities only mention our livelihoods being impacted, but they fail to understand the ecological grief that is attached to our livelihoods being impacted. Given that we necessarily do not care as much for economic gains, especially when they are communal harvests, we should frame this discussion through different lenses that are not normalized within colonial societies.

Yes, our livelihoods are also impacted, but for many of our communities, this is not our first worry. Our first emotion when facing these impacts is ecological grief, and in the climate change discourse, emotions are oftentimes ignored or not acknowledged to the level they should be. Once they start being acknowledged, people might finally understand why our spirituality is also at risk as a result of climate change. This spirituality is also what allows us to conceptualize the relationships we have with our environments as Indigenous peoples because in every Indigenous pueblo of Oaxaca, a god or deity is associated with the milpa. During every harvest, our Indigenous pueblos go into ceremony to ask our god or deity for a healthy harvest. Lita Ser, Ñuu Savi (Mixtec) and founder of Ñaa Ñanga Tijaltepec, shared her ecological grief when her pueblo lost their milpa to the heavy rains.

What does the milpa mean or represent to your pueblo?

The milpa is important to our pueblo because it is the primary source of food—it sustains our life, community, and well-being. Corn, *maíz,* is the main source of food that comes from the milpa and through it, we are able to have other sources of food. For instance, pumpkin, beans, and fava beans. From these food sources, we can start consuming parts of them as they are growing. In the case of pumpkin, we can eat the pumpkin flower. The milpa also connects us to the history of our pueblo, to our first grandparents who took care of us and have allowed the seeds we now have to continue being accessible to us. It connects us to the story of protecting our ancestral foods, our first foods. Everything that comes from the milpa is vital to our traditional foods. For example, corn, *maíz,* is used to make tortillas, our pozole, and even our moles that we consume during festivities. Our history and past is what motivates us to continue caretaking of the milpas. However, it also brings us fear because, for instance, corn has been severely impacted by climate change. During droughts, it does not get enough water and during hurricanes it overfloods our milpas. So there is this fear and caution that we continue to maintain in the back of our minds. Genetically modified crops and other herbicides utilized in nearby agricultural fields worry us as these can severely impact our milpa.

PHOTO 2.1: Lita Ser with families and elders in her pueblo. Photographer: Aldair Garcia Silca

How did the heavy rains impact your pueblo's milpas?

It impacted nearby pueblos as well. I witnessed how many families were impacted by the loss of our food source. The heavy rains destroyed several of our corn stalks and since corn protects the other organisms, once it is damaged, the rest of the vegetables and plants in the milpa lose their protection. It weakened our plants and we lost all of our beans and squash from our milpas.

When the milpas were lost due to the heavy rain, how did that make you feel?

When we lose any type of plant in our milpas, there is a cloud of sadness that takes over me. This is because in my pueblo, we develop and sustain a strong connection with the milpa.

Every time we clean the milpas, sometimes this cleaning means removing some weeds, we talk to the plants. We have an intensive dialogue with the plants. For us, they represent life, not just because they are a food source but also our relatives. During the heavy rains, I heard the tall plants, our *maíz* (corn) fall down. Their roots were losing their connection with the earth because the heavy rains were too powerful for them. It made me feel sad, and our entire pueblo was mourning the loss of our communal harvest, our milpas.

3. The lack of resources offered to Indigenous communities who are already experiencing the impacts of climate change that oftentimes results in displacement

Indigenous communities are already experiencing the impacts of climate change as a result of ecocolonialism, coupled with other human activity that is accelerating climate change. However, ecocolonialism means that Indigenous communities are also not receiving any resources or aid, thus resulting in their displacement. I remember asking my mother, Juana, why she decided to leave her beautiful Oaxacan pueblo to go with my father as he sought refuge in the United States.

Why did you leave Oaxaca?

I left Oaxaca because my parents were struggling with money. My father cultivated mangoes in a small land plot he owned where he also raised some cattle. However, there were droughts that sometimes impacted the mango harvest and the food available for his cattle. This is why my mother decided to sell food in our pueblo to help with the finances in our home. I knew that leaving Oaxaca would allow me to support them more from abroad. I had nine siblings and my parents were getting older so any support I could provide them would really help them.

My grandfather was a small-scale farmer who tried to make a living from his land. Aside from getting the food they would consume from this land plot, he was trying to make small profits from his mango harvest and cattle. During the time my mother was a teenager, they were already facing the impacts of droughts that were reducing the water available for both the cattle and agriculture. Climate change is why Indigenous peoples from Oaxaca and throughout Latin America continue to be displaced. Lita Ser also shared how the loss of the milpas forced some men from her pueblo to leave and try to gain better opportunities in larger cities.

How did the loss of the milpas impact your community?

Aside from starting to worry about where we were going to get our food, some men from families within my pueblo decided to take the bus into larger cities to try to find employment opportunities. The [coronavirus] pandemic really impacted us because most of our families rely on selling our *artesanias,* handmade goods on the streets to tourists in the nearby towns to our pueblo. However, with the pandemic taking place and the closure of many small businesses in the nearby towns, many of our pueblo members were left without a job. Job loss, coupled with the loss of our milpas, our pueblo's food source forced many to leave our pueblo. This means also leaving your pueblo behind, and most importantly your family. Some made plans to not only move to nearby cities but take their journey towards the United States.

Sometimes the displacement that happens among Indigenous peoples is internal as it happens within the cities. This creates the urban Indigenous narrative that is both present within the United States and Latin American countries. They leave their rural communities for a city to ensure they get better opportunities. However, it is more common for many Indigenous people from Mexico and Central America to make their journey to cross the colonial borders to seek refuge in the United States. This makes many of our Indigenous peoples coming into the United States climate refugees.

Neoliberal policies have allowed large corporations to own land in foreign countries. This harms land and Indigenous territories because it is a form of ecocolonialism that continues generating more environmental degradation in Mexico and Central America. This creates a harder environment for Indigenous peoples to thrive in as there is no opportunity to protect their native lands. They cannot compete with larger corporations that have the capital to purchase more lands and generate more profit. Indigenous peoples' livelihoods are jeopardized as they can no longer sustain their communities and families on these smaller land bases that do not favor small-scale farmers. This is one of the driving forces of forced relocation to another country.

Habitat and land loss due to ecocolonialism, coupled with climate change, is one of the most common narratives that Indigenous peoples who have been displaced from Oaxaca and Central America often share. Displacement impacts our Indigeneity because our identities are tied to our land and our ancestral homelands. Whether this displacement is internal, from our rural communities to a city, or external, across colonial borders, it is another way ecocolonialism impacts us. Dr. Daniel Wildcat mentioned that displacement impacts Indigenous peoples *geographically, socially,* and *psychoculturally.* I add that it also impacts us emotionally and spiritually because it manifests as ecological grief. Displacement because of the loss of having our ancestral homelands near or the loss of being able to live on them is ecological grief.

3

Birth of Western Conservation

They [settlers] had the Bibles and we had the land,
natural resources, and precious metals. We closed our eyes
and ended up with their Bibles while they stole our lands,
natural resources, and precious metals.

—MARÍA DE JESÚS, my grandmother

olonization introduced many ideologies that ended up harming Indigenous peoples. We experienced land theft, cultural loss, and genocide. Yet, colonization is not often brought up when we discuss why conservation is needed or why it was birthed as a science field. I believe that in order to do conservation practices, projects, and research justly with Indigenous peoples and communities, scientists who practice conservation need to start by doing these two things:

1. Educate themselves on settler colonialism to understand the root causes of why conservation is now needed.

2. Reflect on their positionality and reeducate themselves on how they were taught to practice conservation practices that

continue to be oppressive toward Indigenous peoples because settler colonialism is deeply embedded in this field.

We often hear the anti-Indigenous statement that the "United States is a nation of immigrants." This statement ignores the colonization the Americas underwent and the genocide Indigenous peoples faced and continue to face under settler colonialism. This statement also creates a utopia that erases the fact that this is indeed a settler country built on stolen lands by stolen people—*slavery*. This is the innocent narrative that is created and manifested inside many classrooms to ensure that white people and settlers' descendants are always comforted for their past sins. It is no surprise why in conservation settler colonialism continues to be ignored as most conservation scientists are trying to undo what their ancestors did or introduced in the Americas during colonization. For Indigenous peoples and scientists, it is important that in the environmental discourse, more dialogues and conversations around settler colonialism and the impacts it continues to have on our environments are mentioned and brought to the forefront.

In my graduate school experience, I started with two white cisgender male faculty advisors for both my master's and doctoral program. However, I started to notice that both advisors were not comfortable with my conversations around settler colonialism and my desire to do conservation projects for my thesis and dissertation that were not rooted in Western science and brought to the forefront Indigenous peoples and their voices. I wanted to do projects that changed the way we are taught to do research in the environmental sciences, including conservation. However, because I brought up racism and settler colonialism, they projected their discomfort as white men onto me. This harmed me and created additional barriers I had to overcome in graduate school as an Indigenous woman. They often referred to me as "hard to mentor," which played a major role in amplifying my imposter syndrome. I started questioning whether I was indeed hard to mentor, and this is just one story, my story. There are countless other stories of Indigenous students who had to face far worse situations and harm enacted toward them by their advisors.

I already knew that academic institutions were not created for Indigenous peoples, so I had to already overcome this barrier to begin with. However, they made it hard for me to be comfortable in my own skin and identity as an outspoken Indigenous woman and scientist. I was once compared to another female Indigenous student who was more timid and quiet. I was told that they had expected me to also be like that student. However, I was not their *norm*—what they had normalized as typical behavior given that they had not mentored many Indigenous students. This is another stereotype that is indeed rooted in settler colonialism, as patriarchy teaches men that they should only be the outspoken ones and advocates. Settler colonialism also grants white people the power and privilege to speak over marginalized communities. This experience taught me that I would be navigating other layers of settler colonialism that would further oppress me and try to silence me as I moved toward the end of my doctoral degree. I was making white men who were also environmental scientists and researchers uncomfortable when I was respectfully questioning why they were doing things the way they were taught. I questioned the teachings they were passing on to me because they were rooted in racism and settler colonialism. This made them uncomfortable, and my experiences show how white scientists ignore settler colonialism because if they start acknowledging it, they have a lot of work to do to confront their white fragility and guilt.

There have been more discussions on racism in the environmental sciences post the antiracism movement led by the Black Lives Matter movement that built momentum in 2020. However, there is much to dissect because unless settler colonialism is acknowledged, the antiracism movement will never get to the root causes of why in the racial caste systems of both the United States, Mexico, and the rest of Latin America, Indigenous and Black people continue to be on the bottom of this hierarchy that makes whiteness its top tier of power and privilege. This dates back to colonization, when both Indigenous and Black people were deemed to be disposable and inferior to whites. Conservation is not a field that is free from subjectivity as it was a field founded by white people, in particular men, to practice their colonial notions toward nature, our environments, and Indigenous peoples.

Conservation is a Western construct that was created as a result of settlers overexploiting Indigenous lands, natural resources, and depleting entire ecosystems. According to *National Geographic*, conservation is defined as "the care and protection of . . . resources so that they can persist for future generations. It includes maintaining diversity of species, genes, and ecosystems, as well as functions of the environment, such as nutrient cycling."[1] Conservation did not exist precolonization because Indigenous peoples viewed land as communal, meaning no one person owned it. Land was also a part of their identity, *Indigeneity*. It is important to mention that these values continue to be sustained among many Indigenous communities, and this is why they still have communal food sources and land. I was taught that our lands do not owe us anything. They are not here to feed or shelter us, as our lands were not created to serve us. As I mentioned before, many of our creation stories, how our Indigenous ancestors came into existence, point to them being made from earth, our lands. This means that our lands are a part of who we are and in order to respect and thank them for giving us life we have to protect them.

However, during colonization, everything changed, as settlers stole our lands and made them their property while leaving us out of these discourses because they had the notion that we would go extinct or completely assimilate. This is why they gave us our Bibles as they stole our lands, natural resources, and precious metals as my grandmother always told me.

As my grandmother mentioned, settlers came for our land, natural resources, and precious metals; thus they stole our land and later commodified it and made it their property. Conservation, in my view, aims to undo some of the harm settlers generated, but many conservation initiatives continue to fail to uplift and center Indigenous peoples. Indigenous peoples always knew to take only what they needed, and they continue to uphold some of these values today. Taking more than what is needed is deemed as bad medicine in my Zapotec community because greed pulls us toward evil spirits and thoughts. Greediness also angers our gods and deities, and when they get angry they end up punishing our community, not just the greedy person.

We also have to ask our natural resources, both plant and animal relatives, for permission before we extract, hunt, or remove them from their environment. For example, when I remove a plant from its location, I ask its spirit for permission, pray before I remove it, and leave medicine in its place. This medicine consists of copal, lavender, and a medicinal plant native to the region where I am removing this plant. For example, if I am removing a plant in the state of Washington, I often incorporate cedar into the gift I leave behind in its place. Indigenous peoples have built and sustained reciprocal relationships with their plant and animal relatives. If something is taken, something is left in return. Given that it is a spirit we are removing, we must leave part of our spirits in its place. This is why the gifts we leave behind when we remove the plants are part of our traditional medicines, like copal is for my Zapotec community. This is the mentality that allowed our lands to be replenished with natural resources.

All of that changed during colonization and postcolonization. Land is considered our mother for many Indigenous communities, including my Zapotec and Maya Ch'orti' nations. Thus, it never had a monetary or economic value like what was imposed by settlers and colonizers. It is hard to envision selling or owning our biological mother as humans, and this is the same way we view our Mother Earth. However, colonization introduced and created systems that were driven by capitalism and generating commonwealth for colonizer countries and descendants across the Americas. This mentality and way of viewing nature as a commodity and land as a property created *boom-bust cycles* that impacted both the native flora and fauna ecosystems of the Americas. Boom-bust cycles refer to the rapid decline of species due to over extraction, overharvest, and over-hunting.[2] This is the main reason why the population size of a lot of native and cultural keystone species severely declined. Coupled with deforestation and other land loss, many animal and plant species were severely threatened and are now endangered. For Oaxaca, these boom-bust cycles that resulted in the overhunting of many wildlife species and destruction of their habitats led to the rapid decline of many wildcats native to the region of Oaxaca and other parts of Latin America. These wildcat species include jaguars (*Panthera onca*), margays (*Leopardus wiedii*), and ocelots (*Leopardus pardalis*), among others. The jaguar is a cultural keystone to the Zapotec people. This is why the Zapotec deity Cocijo, god of rain and lightning, had the facial characteristics of the *beedxe'* (jaguar) to represent our earth. This god, like many others that are vital to the Zapotec people and culture, represents parts of their natural elements and plant and animal relatives.

This is how important and integral natural elements and resources are to many Indigenous peoples across Latin America. All gods and deities are known to control the natural elements such as the sun, rain, and lightning, among others. Natural elements are what nourish all natural resources. It was a spiritual cycle that was created between humans, natural resources, and gods and deities (spiritual and powerful beings). This spiritual cycle represents the strong connection and holistic view of nature Indigenous people continue to sustain. Everything is interconnected from outer space (e.g., moon, stars, etc.) to the bottom of our oceans. However, Western knowledge and way of thinking have separated many of our landscapes into plots of land, and this is clearly seen throughout conservation initiatives and projects. The establishment of colonial borders and borders within several nature preserves continues to separate entire landscapes into smaller plots of land.

Native animals to the region play an important role in Indigenous ancestral history. These native and important animals are what I refer to as keystone cultural species. Without them, a component of our Indigenous ancestral history is lost. This is why the jaguar, among other wildlife native to Oaxaca and across Latin America, are important to Indigenous communities. If the jaguar were to become extinct, the spirits that guard the god Cocijo will also become extinct. This is a way Indigenous peoples continue to experience cultural loss. When animals or plant species become endangered or threatened, Indigenous cultures are also threatened.

Cocijo has been present among the Zapotec culture since time immemorial, or as anthropologists have coined, since the *pre-Columbian era*. Cocijo, among other deities and gods, has been found in many Zapotec *codices* (glyphs), written records found on archaeological sites. Unfortunately, during colonization, our history was erased and burned by colonizers and settlers. They wanted to destroy our identity because they wanted to bring forward their racial caste and other systems to steal our lands and exploit our natural resources. We lost many of the relationships our ancestors had with our deities and gods. Now we are reclaiming those histories and reconnecting with our ancestral spirituality that colonizers burned so that we could forget it and they could eradicate it. Our codices were burned, and through this action colonizers burned the historical written records our ancestors were maintaining for our cultures.[3] This is why settler colonialism is important for conservationists to also start discussing and acknowledging. It was a genocide that no settler country (e.g., United States or Mexico) wants to acknowledge because they want to continue portraying their innocent history of how the Americas were founded and built.

When discussing conservation it is important to mention that *conservation* was founded as a movement to preserve wilderness, wildlife, and other natural resources. Given that it is not an Indigenous way of life, because it was not needed before colonization, many Indigenous languages do not have a word that translates directly to *conservation*. In the Zapotec variant that my family speaks, the closest word that we could think of that may refer to conservation is *gapanú* or *rianda. Gapanú* means to take care of something and *rianda* means to heal. However, these two words do not embody what conservation means to its extent. Thus Indigenous languages can serve as indicators of what has been introduced into our lands. This is how we know for a fact that conservation indeed did not exist among our communities precolonization.

However, our languages also adapt with modernization, and our vocabulary continues to grow as well. For my Zapotec people, *taking care of* or *healing* are the closest words that come to mind when we try to explain conservation in our Indigenous languages. This is why I opt to use the word *healing* as opposed to *conservation*, as conservation is not really a part of the identity or cultural values I uphold as an Indigenous woman. For me, *healing* or *caring* are more aligned with my culture and Indigeneity.

National Parks

National parks were created for conservation purposes and are managed by government entities.[4] In the United States, the movement that created national parks and conservation was led by John Muir, Gifford Pinchot, Theodore Roosevelt, and George Bird Grinnell, among others. Given the list of those who are known as the founders of conservation, we can indeed bring forward the fact that the field of conservation was created by white men. Thus, national parks' frameworks and paradigms are grounded in beliefs that are rooted in settler colonialism, patriarchy, and capitalism. These were the paradigms and frameworks that were uplifted and instilled in this country during this time period as settlers continue urbanizing and creating settlements throughout the United States.

Conservation gained more attention and traction in the United States during the establishment and creations of these national parks. National parks were created under the notion that it will preserve nature's beauty for all—the common person who often felt unheard or unseen during these times. However, national parks hold on to a dark history of further displacement, dispossession, and removal of Indigenous peoples. It became a part of the American family legacy, as national parks were created for family gatherings, celebrations, camping, and other family events. Glob-ally, *conservation* was coined by President Roosevelt, as he made this a modern idea that was rooted in the United States' policies and acts during this time. Conservation became a movement to avoid western expansion and urbanization by creating sites where no federal or private sector could do anything without proper authorization from the US government. This of course also included displacing Indigenous peoples from their ances-tral lands if their lands were deemed as national parks–worthy.

Western conservation continues to be a movement that is supported and validated by strong scientific evidence and studies. The strong scien-tific evidence and studies continue to ignore Indigenous ways of knowing and knowledge systems, as Indigenous knowledge is not deemed worthy enough to be considered science. This demonstrates how science, in this case environmental science, continues to be subjective rather than objec-tive because it is rooted in Western European philosophies and knowledge. It is interesting that scientists and many others continue to treat science as an objective approach, when in reality, it is rooted in settler colonialism and cultural values. This is why the sciences continue to dismiss Indig-enous peoples and other communities of color, especially those who are marginalized under whiteness. It means that it will be common for an Indigenous student to lack a sense of belonging in the sciences unless we adapt and assimilate to how science is perceived and the cultural beliefs and philosophies it upholds.

This is why I say that being Indigenous does not equate to doing Indigenous science, because Western science is not Indigenous science. Western science is subjective and based on Western European philoso-phies and histories that continue to credit white European men as the founders of many scientific disciplines. For example, Ferdinand Magellan is considered the pioneer in ocean sailing. He was Portuguese and was sup-ported and aided by the Spanish monarchy that colonized most parts of Latin America. However, many scholars have concluded and shown strong evidence that it was Polynesians, Indigenous people of the Pacific Islands, who had navigated the oceans and had completed ocean voyages precolonization.[5] The history of Western sciences has been whitewashed, meaning that Indigenous contributions to science are ignored, suppressed, and not acknowledged. Only contributions and discoveries by white European men are what are praised and referred to as the foundation of the sciences. This is why it is important to always question science because in its roots lie years of oppression, racism, and dismissal of Indigenous cultures, communities, and people. Yes, science is valid and has helped solve many problems such as the pandemic of 2020, but this does not mean it is free from racism, oppression, or subjectivity. This deeply applies to conservation, as it is rooted in the Western environmental sciences.

Conservation continues to function on five major pillars that I have identified throughout my academic journey, scholarship, and experiences in the environmental discourse. These pillars are

Pillar 1: To maintain essential ecological processes;

Pillar 2: To maintain the ecosystem's living resources;

Pillar 3: To preserve genetic biodiversity;

Pillar 4: To yield sustainable species and ecosystems for humans; and

Pillar 5: To protect wildlife from both human-caused and environmental harm.

These five pillars demonstrate that Western conservation focuses on the usefulness of nature to humans, as natural resources hold economic and historical values in the society the United States settler governments created. It creates a utilitarian use of nature and does not focus on what humans also have to offer nature in return. Western conservation is blind to cultural values of natural resources—especially those important to Indigenous peoples. For Indigenous peoples, we never associate a value to our environments. These values range from economic or historical value to also this notion of beauty. Beauty is also a value that established the notion of pristine wilderness and purity because European beauty focuses on this.

The conservation initiatives and paradigms the National Park Service (NPS) generated were also amplified by many naturalists including John Muir. His writing about the natural environment he witnessed romanticized the notion of conservation and what it meant "to conserve the scenery and the natural and historic objects and wildlife."[6] This phrase was one of the main goals of the Organic Act of 1916 that defined the NPS conservation mission. It continued to embed this notion that in order to preserve a natural landscape, humans were not a major component of this conservation paradigm or framework. National parks became places where humans could enjoy the beauty and some recreational activities, but national parks were not meant to become places where humans could also contribute to the healing of these landscapes. Landscapes across the Americas need healing because they have experienced harsh environmental impacts as a result of urbanization, industrialization, climate change, and other challenges. As a result, national parks became known as an American invention, as the main reason why they were established was to protect the western parts of the United States (i.e., Yellowstone, Yosemite, etc.) for tourism. Tourism is oftentimes conflated with conservation through ecotourism. While it indeed supports and benefits certain landscapes and species, this ecotourism is deeply rooted in the displacement and removal of Indigenous peoples. Of course, this was a goal that became a priority among NPS conservation frameworks and paradigms once Indigenous peoples were removed from the lands. It never included the local and Indigenous peoples of areas that were designated and converted into national parks throughout the Americas.

The conservation frameworks and paradigms the NPS follows—that separate humans from nature—are one of the reasons why the creation of national parks led to impoverished conditions for the local communities. National parks were established on the displacement of Indigenous peoples, and since 1916, when the first national park was created, not much has changed. On the contrary, the impacts it has on rural and Indigenous communities is one of the new advocacies many individuals are speaking out against—even the World Park Congress.

Montecristo National Park, El Salvador

Metapán is a beautiful place as it is very green, home to
many animals, and our sacred waterfalls. I learned how to swim
in the river that goes through there my ancestral lands.

—MY FATHER, Victor

In El Salvador, one of the national parks that has high ecotourism is Montecristo National Park. This is a national park near the area where my father grew up in Metapán, a Maya Ch'orti' ancestral territory. During the colonial periods, Spanish settlers created a huge cattle ranch they called Hacienda San Jose. Unfortunately, this cattle ranch contributed to the large deforestation of the rainforest and in the 1710s the lands were further exploited for iron mining. My father recalls the beautiful lands of this rainforest that was purchased by El Salvador's government in 1971. The Salvadoran government was able to purchase over two thousand acres of land that they wanted to convert into a national park.

It was interesting to see the influence of white men in the decision of converting this rainforest into a national park as Marc L. Rocher, a French man, and Howard E. Daugherty, a United States scientist, advocated for the Salvadoran government to use this land as a conservation-designated area. Rocher advocated for the lands to be converted into an agroforestry area and Daugherty advocated for it to be converted into a national park, a US-based phenomenon I have discussed previously in this chapter.[7] In order to avoid any conflicts, El Salvador decided to claim that there were no Indigenous peoples living in this area anymore. However, during this time, the civil war was starting in El Salvador, so there was no time to fight over land as many peasants and Indigenous peoples already knew the government was further oppressing them.

When El Salvador purchased this land in 1971, there was still 43 percent forest covering. In 1987, a few years before the civil war ended in El Salvador, it was converted into a national park. Throughout this history, there were about one hundred families that lived in the area that were descendants of Indigenous and Black people who were forced to work the cattle ranch during colonial times. However, the government decided to ignore them and fight against them to ensure they did not oppose the creation of the national park. However, since this happened during the civil war and we already know that the government was using violent military tactics against anyone who wanted to speak up against them, the families were not able to do much.

However, these are Maya Ch'orti' ancestral lands that had gone through severe ecocolonization when they were converted into a large cattle ranch and later exploited for iron. For me personally, as a Maya Ch'orti' woman who also has strong roots in El Salvador, it is hard to talk about the Indigenous history of my paternal ancestral lands because the Indigenous population represented over 51.9 percent of the population between 1769 and 1798, and most recently in 2020 our Indigenous population is only 10 percent of the country.[8] This is due to the civil war that targeted Indigenous peoples first and the ongoing impacts our Indigenous people continue to face in Central America that has resulted in forced displacement. We will discuss this more in chapter 5.

The Montecristo National Park buried the Indigenous legacy and history that was made up of mostly Maya Ch'orti' people because it was stolen from our Indigenous ancestors and converted into a large cattle ranch during the colonial era and later further exploited for iron mining. During the time period when El Salvador was converting it into a national park, a civil war was on the rise and the government was using forceful and violent military tactics to control any Indigenous person or community that was trying to revolt or fight against them. This national park buries a dark history that resulted in the genocide against my community of Maya Ch'orti' people, and it continues to serve as a reminder of how national parks were created to do just that: displace and bury the Indigenous history of those parks. Today, this park generates a large ecotourism revenue to the point that a walkway and a hiking trail were built so that they can go through the park. It is also no surprise that two white men, Rocher and Daugherty, were the ones giving advice to the Salvadoran government on what kind of conservation approaches they should use instead of asking the local and Indigenous communities of this country.

Top-Down Approach

Conservation continues to teach scientists that scientific knowledge is more valuable than Indigenous knowledge, and therefore, science knows what is best for Indigenous communities. Science continues to follow a top-down approach, and as long as it does this, Indigenous communities will continue to be oppressed under this framework. This top-down approach centers scientists and conservationists as the knowledge holders who continue to come into Indigenous communities and territories and advise Indigenous peoples what to do with their environment. This top-down approach also fails to recognize Indigenous science and knowledge systems are essential because attached to them are ancestral knowledge and lived experiences within that environment. This approach introduces demands, practices, and commands that oftentimes come in conflict with Indigenous communities' way of life. This way of life might have adapted to modern times, but it has been sustained among Indigenous communities since time immemorial. Conservationists often fail to recognize that many Indigenous environments are endangered or in constant threat because of the practices and beliefs of settler colonialism introduced to the Americas and not because of the Indigenous communities themselves. Extractive industries such as oil, coal, and iron, among others, continue to create negative environmental impacts on our landscapes. These methods were introduced during colonization as settlers and colonizers were interested in finding precious metals and riches outside of their European countries for their respective monarchies. However, most scientific fields continue to try to separate themselves from the legacy of settler colonialism because it is an uncomfortable conversation for most scientists.

According to the National Science Foundation, in 2017, 71.1 percent of doctoral degrees in the sciences were awarded to whites in comparison to the 0.4 percent awarded to Indigenous students.[9] Thus the uncomfortableness of these conversations that acknowledge how settler colonialism is rooted in the sciences impacts mostly whites. However, it is important to realize that for Indigenous peoples, settler colonialism and its impacts continue to be our everyday experiences. While white scientists can choose to ignore these conversations, as Indigenous peoples we are reminded every day of how our culture, identity, lands, and other parts of our lives continue to be threatened and impacted. We see how white scientists continue to be oblivious to settler colonialism and how deeply rooted it is in the environmental sciences, physics, medicine, and other science fields. There is a failure to reflect on the founding history of these fields and how these founding histories continue to play a major role within the fields and disciplines that have been created from within.

Settler colonialism grants certain scientists from wealthy countries such as the United States permission to go to other impoverished countries throughout Latin America and create their own research projects, centers, and other endeavors while further displacing the Indigenous peoples of those areas. Ecological and conservation research is often conducted by scientists from the United States, Canada, and other European countries that have the resources and autonomy to decide where they want to do research. In higher academia, we are taught that we can create a research grant proposal for "anywhere in the world." I have come to learn that most of the time this statement refers to impoverished countries, and in the Americas that is Mexico and Central and South America (Latin America). This perpetuates the cycle of *helicopter research* where researchers from wealthy countries go to an impoverished country, conduct their research studies, and then return to their countries to analyze the data they collected and publish it, oftentimes not even including or consulting the local people of those countries. In Latin America, oftentimes this helicopter research leads to having white and Westerner researchers from countries that have a long history of colonization write our stories instead of supporting Indigenous peoples so that we can write our own stories.

I remember my visits to Oaxaca when my grandmother was still alive. Oftentimes I would go to the local shops or outdoor markets to help her purchase fresh fish and other produce. We would sometimes see white men and women with fancy cameras and equipment coming down from their trucks. My grandmother would always roll her eyes and tell me to keep on walking and not to engage or talk to any of them. I did not understand why she despised them, because I thought they were journalists or newscasters, similar to what we had in the United States. Oftentimes many of the people in our pueblo ignored them and their Spanish interpreter as well. Under her breath my grandmother whispered to me, "*M'ija*, these are people similar to anthropologists. They are here to collect our stories and statements because they say they are conducting 'research.' However, they have offered so many people stipends for their stories and interviews but did not pay them anything. On top of that, they are working on a book to write our stories. What do you think about that?" At such a young age I replied to my grandmother, "Why don't they help everyone in the pueblo instead to learn how to read and write so that our people could write their own stories instead?"

Obviously, researchers who conduct helicopter research are not interested in what they can offer or what the community might benefit from, as their main goal is to collect data and then publish it to advance their careers. Helicopter research is the most common form of this top-down approach that the sciences continue to teach and amplify in academic institutions. Determining what kind of research is helpful without consulting the community or asking them what might benefit them is a top-down approach that can further harm the community, especially Indigenous peoples. In conservation, scientists are also taught that if something worked in one country, it might work in another one. This creates this mentality of *one-size-fits-all.*

PHOTO 3.1: My grandmother and grandfather. Family Archive.
Photographer: Unknown

Utilizing the top-down approach promotes the creation of one-size-fits-all conservation solutions and practices that may not necessarily work for all Indigenous communities because Indigenous communities are not monolithic and their way of life is place based. Given that coastal communities are different from inland communities, the same conservation approaches will not work as they have to be adapted to meet the community's needs. This is why it is best to center the community first, which derives from the opposite spectrum, following the bottom-up approach.

I recall in my graduate school that a professor was very mad that his potential research project was cancelled because the local federally recognized tribes of the state of Washington were not interested in creating marine protected areas (MPAs) in the Puget Sound and Salish Sea. For him this was a great way to protect salmon because he had done other research projects in other developing countries to create MPAs for conservation purposes. MPAs are the conservation frameworks that have been applied to many impoverished and global south countries. He mentioned how he had "wasted" a lot of money trying to get this project started just for the tribes to say no to his proposal. This is an example of the combination of different top-down approaches I have discussed. He thought he knew what was best for the community as opposed to the communities, in this case tribes, knowing what was best for them. This is the settler colonialism that is embedded in conservation, where non-Indigenous scientists have not developed the same relationships with the local environment as tribes who have been cherishing these relationships for generations. He wanted to apply the one-size-fits-all model to tribes in the state of Washington because he had done similar projects on MPAs with other Indigenous communities in other countries.

In conservation, we are taught that practices and approaches that were *successful* should be what we apply to different regions, places, and communities. This tends to ignore that every community holds a different set of values and relationships with their environments. Also the success of conservation practices and approaches is often determined by the scientists, not the local communities. Lastly, he had already decided his *research* project, because he was the scientist, instead of asking the tribes what kind of conservation projects they wanted to do in regard to their marine resources.

We need to start discussing this top-down approach that is embedded in the sciences, in particular anything related to our environment, where scientists believe their academic credentials and experience can outweigh lived experiences and local knowledge. This is why reversing this top-down approach to become a bottom-up approach is crucial and essential to benefit local environments and communities.

Bottom-Up Approach

The bottom-up approach allows conservation initiatives and projects to come from within the community. Contrary to the top-down approach, it centers the communities from the start as it does not assume or introduce demands, commands, and practices that are oftentimes given without building rapport or having experience living within the environments scientists are trying to protect and conserve.

I have practiced this bottom-up approach throughout my graduate career, and I think that given that I understand from firsthand experience the harm research can do toward Indigenous communities, I have become mindful of how I, too, practice research. For my master's thesis, I had built relationships with the local Coast Salish peoples, and I started asking them what kind of project would benefit them. They discussed coding environmental justice cases so that people in the state became more aware of the environmental injustices they were facing. This is what led me to completing my thesis project on coding environmental justice cases they reported to me. These cases allowed me to critique how environmental justice is denoted in policies that continue to dismiss the cultural values that are important for Coast Salish peoples. For my dissertation, I also did the same thing, but in this case with the urban Indigenous community of Seattle. My strong relationship and immersion in the urban Indigenous community allowed me to ask them what kind of environmental project they could benefit from. This led me to the Daybreak Star Indian Cultural Center.

The bottom-up approach is also an important step when conducting community-based participatory research (CBPR). CBPR is defined as a collaborative approach to research, especially in the health fields that tend to work mostly with communities. CBPR can become congruent with Indigenous communities because it attempts to dismantle some of the impacts of research and settler colonialism. It allows Indigenous peoples to serve as the leaders and consumers of the research projects meant to benefit their communities rather than just serving as research subjects. It also allows for the creation of an effective collaboration and destruction of power differentials between the researchers, community members, and relevant organizations.

CBPR moves away from the top-down approach Western research follows toward a bottom-up approach that some Indigenous communities implement in their governance. By positioning research into this bottom-up approach, research ends up demanding less work from Indigenous peoples and empowers them to reclaim their autonomy. CBPR allows us to center and uplift the communities' voices despite having to work with some organizations that continue to follow the Western scientific approach.

However, it is important to understand that CBPR is still conditioned through Western education and research because CBPR still allows for the integration of Western research methods, which can create an Indigenous context and not necessarily follow Indigenous methods. Since CBPR can also be practiced and implemented by non-Indigenous peoples, an Indigenous context is created; non-Indigenous peoples become the voice for the Indigenous community they are working with and speak on behalf of them. In order for CBPR to be modified for Indigenous communities, I identified six principles during my doctoral degree to incorporate into CBPR. These principles were identified through my extensive trajectory in conducting community-based work that oftentimes was not done for academic or research purposes.

1. **Follow and create fluid and dynamic approaches that do not follow the linear research method.** This means removing your work from always following the scientific method. It is important to ensure that Indigenous communities and people are not treated like test subjects, because this is a form of colonization that has been used against Indigenous communities. They should be informed of every step and be allowed

to sit at the table where decisions are being made, not just informed of the decisions once they are made.

2. **Respect tribal sovereignty and Indigenous autonomy.** It is up to the Indigenous communities to decide what research and work are allowed on their lands and with their communities. In the example of the professor who wanted to introduce MPAs in the state of Washington, he had failed to understand that tribal sovereignty is important for the tribes as it granted them the right to fish salmon and other aquatic native species to the state. Creating an MPA would limit the rights they fought hard to obtain through the Boldt Decision. The Boldt Decision made the state of Washington respect the treaties they signed with tribes and allocated 50 percent of all fish stock (e.g., salmon, shellfish, etc.) to the tribes.[10]

3. **Follow Indigenous protocols and their way of being and doing things in their communities.** Western science and research do not always align with Indigenous ways of life; therefore, it is important to listen to the approaches and how they want things done in their communities. I often hear from researchers that they did not receive enough community engagement in their projects, and to me this points to the reality that their scientific approaches were not aligned with the communities' way of being or doing certain things.

4. **Respect intellectual property.** In academia, we are pushed to publish, and this has created a lot of Indigenous scholarship written by non-Indigenous peoples. If an Indigenous community shares important information or sacred knowledge, it does not become the researcher's intellectual property, especially if they do not belong to the community. This leads to co-option and theft of Indigenous knowledge.

5. **Embrace all Indigenous epistemologies relevant to the community.** Western knowledge is not Indigenous knowledge. This means that things are thought through differently by Indigenous communities. An example of this for my Zapotec community is that in order to do any initiative or project in our community, I must first get permission from my elders, in particular our matriarchs. This means that I have to sit within an elder's circle and listen to what they have to say. Time is nonexistent in our communities and oftentimes these meetings cannot occur over just one or two hours. They can take days or weeks, and researchers must be open to this and understand that the Western notion of time is sometimes something our communities do not prioritize.

6. **Be an Indigenous-led project.** Researchers need to reflect on why they are leading this project. Based on the researcher's positionality, it is okay to step back and do the behind-the-scenes work to create an Indigenous-led project. Conservation and the environmental sciences continue to romanticize Indigenous peoples and work with Indigenous communities. However, it is important to sometimes step back and let the project be led by the community. We know that scientists and conservationists have resources that are oftentimes not provided or available to the Indigenous community. This is what a true ally does for our communities. They provide resources and support for the community in the form of a consultation and let the community take the lead on the project. I hear white or non-Indigenous scientists oftentimes mention that they are making an Indigenous person from the community they worked with a coauthor on their paper. My question is always, Why was this paper not written entirely by the community? To me this feels like a

way to ignore the reality that some projects are meant to be entirely Indigenous-led with white and non-Indigenous scientists supporting along the way.

Incorporating these principles modifies CBPR core principles and makes them more relevant to Indigenous community-led research. By doing so, CBPR is conditioned away from colonial Western education and facilitating a project that centers and amplifies Indigenous voices. This allows us to move away from doing research on Indigenous communities toward research led by Indigenous peoples.

Ecological Noble Savage

I have heard throughout my environmental science trajectory the following statement: "I want to gain experience working with Indigenous communities." This statement has been made by white and non-Indigenous scientists and professionals throughout my life. It is a bit weird that working with Indigenous communities and gaining this experience serve as a way to amplify non-Indigenous scientists' careers, and it always makes me feel uneasy. Indigenous communities should not be an academic or professional goal for anyone. This is the settler colonization that is still embedded within conservation and the environmental sciences. It is also important to note that this desire and goal to work with Indigenous communities is rooted in the stereotype of the ecological noble savage. When discussing the relationships Indigenous peoples have with their environments, I always reflect on this stereotype because it is important to acknowledge that this stereotype does indeed exist.

As an Indigenous woman who pursued her career in the environmental sciences, this stereotype has shadowed my professional experiences. I think the statement that has negatively impacted me the most was "You are an Indigenous cultures expert," which I've been told by several white people. I think making people experts on the identities they embody is only common for those who embody marginalized identities. Also, this treats Indigenous cultures and people as some sort of expertise, when in reality it is an identity that needs to be respected and not treated like a research subject, as anthropologists did during the colonial time period. I have also been asked in several of my environmental classes to give the Indigenous experience, when in reality we are not monolithic and I cannot speak for all Indigenous peoples. In this book, I provide examples that include my Indigenous communities and those who I have a strong relationship with, but this does not mean that every example should be applied to all Indigenous peoples in general. That causes harm as it fails to recognize that we are all different and embody different intersectionalities as Indigenous peoples.

I have often felt romanticized by non-Indigenous peoples because the ecological noble savage stereotype has been deeply integrated in the environmental discourse. This stereotype dates back to colonization, when Europeans viewed our ancestors as innocent and mystical creatures that were free of sin. Given the stewardship and care we gave to our lands, the nature that surrounded the New World, as they called it, was different from their European lands because ecocolonialism had not severely altered our landscapes or destroyed them yet like they had done back in Europe. It led them to realize that our lands in the Americas were replete with natural resources because our ancestors' way of life always told them to only take what was needed. Consumerism and capitalism were not ideologies our ancestors carried or believed in. We had a strong relationship with nature, as we never viewed ourselves separate from it. This view never motivated our ancestors to extract from our environment, as the environment was also a part of us, our lives, our culture, and our spirits.

However, we have to be cognizant that today not all Indigenous peoples have a strong relationship with nature or their environments because of displacement and ecocolonialism, which ultimately fractured our relationships with nature and on some occasions has completely disconnected us from nature. When speaking about Indigenous peoples and their kinships with Mother Earth, it is important to become aware of this stereotype and ensure that the environmental narratives and discourses do not continue to amplify it. This will allow us to start addressing those colonial legacies that are still embedded in the environmental discourse.

These legends and myths that surround the stereotype of the ecological noble savage are the same reasons why as Indigenous peoples we are either romanticized or oppressed. These legends and myths that are rooted in Western religion were also used as tools to justify what colonizers did to our environments and landscapes. It also justified our genocide in their eyes, under Christianity. The romanticization of Indigenous peoples through this stereotype of the ecological noble savage also uplifts and centers Pan-Indigeneity. As Indigenous peoples, we were seen as one monolithic group during colonization and this same harmful rhetoric is still recycled among many today.

Pan-Indigeneity is what conflates Indigenous peoples as monolithic groups, despite having different experiences, perspectives, voices, and intersectionalities. As mentioned before, our relationships with our environments and nature are different because our environments are different. We cannot expect a coastal Indigenous person to be similar to an Indigenous person from a desert area. This Pan-Indigeneity is harmful because it also continues to bring to mind people from the past as opposed to people who are also contemporary and have adapted like many communities to modern times. It assumes that Indigenous peoples living today are more like their ancestors who were living in the past, thus not recognizing modern and current lived experiences.

It is unfortunate that Pan-Indigeneity and the stereotype of the ecological noble savage are present among many environmental groups or organizations today. It is visible on job descriptions for these environmental groups that are moving to become more inclusive where they make statements like "seek and engage in nation-to-nation collaborations working with the native peoples of the [area]."[11] Indigenous people know that these are jobs that they do not qualify for because educational and work experiences have often been denied to Indigenous communities for years. This type of discourse utilized in job descriptions continues to perpetuate that working with Indigenous peoples is an academic or professional experience scientists can seek, but it does not accept or promote lived experiences, which is the type of experience Indigenous peoples embody. It is a subtle form of settler colonialism that is embedded in the environmental sciences as many jobs that end up going to white people or non-Indigenous people ask for this qualification, "experience working with Indigenous peoples," but ask for countless other requirements and qualifications that we know Indigenous peoples do not have because they were never given the opportunity to pursue such opportunities.

Racism in the History of Conservation

Like most movements and concepts, conservation has stayed within the time period it was created. Until most recently, the racism these founders of conservation believed in and enacted against Black and Indigenous communities is finally being called out and brought to the forefront. For instance, many of these men who are considered the founders of conservation were anti-Black and anti-Indigenous, and they verbally expressed this very freely as it was allowed during this time period. This is why it is no surprise that the field of conservation has not been inclusive to many people of color, in particular Black and Indigenous communities. These beliefs continue to be within its core as the founding stories and are why conservation and how it continues to be deemed today continue to uplift and benefit white men while dismissing Indigenous peoples. It is no surprise, then, that many conservation organizations today continue to be led by white men and they are not diverse workplaces. The irony is that conservation is portrayed as something that is supposed to serve all communities and protect the environment for everyone. Yet, the people making the decisions in regard to conservation and the management of our natural resources are not diverse. Until these beliefs are called out, they will remain at the core of conservation.

This racism and exclusion of Indigenous peoples are also portrayed in the names many monuments and sacred sites were given within many national parks. For example, the history of Yellowstone National Park shows us how racist histories continue to be celebrated. During the 1870s and 1880s, when Yellowstone National Park was created, the Native American tribes of this region were forcibly removed and relocated. This marked the continued displacement Indigenous peoples have faced in the United States since settlers first arrived in the Americas. These national parks are the ancestral lands of many Indigenous communities, yet the Indigenous communities of those lands are ignored from the national park discourse. This is why Indigenous peoples have fought for their cultural revitalization and their right to practice their ceremonies within these national parks. For many Indigenous communities, their relationships between national parks and their spirituality are sustained through practices that continue to take place at the parks, as these are their sacred sites.

There is historical evidence that the Shoshone, Crow, Arapaho, Northern Cheyenne, Blackfeet, Flathead, and Nez Perce peoples spent time on the land before it was converted to what is now known as Yellowstone National Park. Archaeological evidence has found obsidian tools that belonged to these tribes. However, there is still no recognition on the national park's behalf to admit how its founding history severely impacted Indigenous peoples, the original stewards of these landscapes.

It is also interesting that when national parks were founded, the United States passed policies to preserve Indigenous history but without the input of Indigenous peoples. Following Yellowstone's establishment, President Theodore Roosevelt signed the Antiquities Act in 1906. This act gave presidents the authority to create additional national monuments to preserve areas of natural or historic interest on public lands, largely to protect prehistoric Native American ruins and artifacts. While there was a desire to protect Native American ruins, there was no desire to protect Indigenous peoples and their cultures as they were denied access to national parks during these times. This was also due to the violent assimilation tactics the United States government was using against Indigenous peoples. Their priority was to create a natural wilderness with no nature-culture nexus that interconnects humans with nature, primarily Indigenous peoples. When they were displaced, they could no longer use the natural parks for their sustenance, cultural practices, and spiritual traditions.

Yellowstone not only marked the forced removal of Indigenous peoples, it also celebrated the genocide enacted against Indigenous peoples during these times by naming many monuments after violent people that led these atrocities. For instance, Mount Doane, located in Yellowstone Park, is named after US Army Lieutenant Gustavus Doane. Doane was the leader of the massacre of over two hundred Piikani people. Naming something after a person who led such violent massacres against a tribe is the same as celebrating genocide. Yet, in most national parks, monuments are named after such leaders. Not only did the displacement of Indigenous peoples threaten their cultural survival, the historical figures who threatened them are celebrated on their ancestral lands. This is another way settler colonialism manifests through national parks, a conservation practice.

Not all history is meant to be celebrated but only remembered so that everyone can learn from past mistakes that led to genocidal acts. However, behind the histories of national parks, conservation, and the environmental sciences, there are never any Indigenous voices or people celebrated. On the contrary, their oppressors are the ones celebrated. We are seeing some changes to this, but there is a lot that needs to take place to undo the celebration of hurtful histories.

Given the antiracist movement sparked during the 2020 pandemic, we are witnessing major conservation organizations that were founded by the same violent men start to unravel and dissect their racist histories. For example, the Sierra Club is finally facing and admitting that their founder, John Muir, referred to Native Americans as "dirty" and utilized the colonial mentality to rename the Sierra Nevadas despite it already having a name given by the Ahwahneechee people. On top of this, he made racist remarks that diminished how Indigenous communities also lived and continued to thrive after the genocide and colonialism our ancestors faced. This is why I can say that conservation is an environmental movement that is still rooted in settler colonialism and continues to uplift racism. Until there is more unraveling and reflection on the racist beliefs and actions that founded this field, it will continue to recycle these norms and beliefs that will continue to further harm communities of color.

I think an important question to ask ourselves is, Why is the face of conservation still white men? From my understanding, Indigenous peoples took care of their environments, landscapes, and nature harmoniously before settlers arrived. The birth of Western conservation is rooted in racism and anti-Indigeneity; thus this is a call for conservationists and environmental scientists to start questioning whose and which histories are celebrated in their environmental organizations, nonprofits, and within national parks.

4

Indigenous Science:
Indigenous Stewardship and
Management of Lands

When colonizers burned our histories or trapped it behind
museum glass cases, they also burned and trapped generations of
traditional and ancestral knowledge.

—MARÍA DE JESÚS, my grandmother

I ndigenous communities of the Americas had advanced scientific
knowledge that allowed them to build complex structures that continue to amaze many today. Some of these structures include the pyramids and are still studied and referenced in many scientific disciplines.
Yet Indigenous peoples are never credited for their immense and rigorous
scientific knowledge that was fractured and erased during colonization.
Many of these scientific fields have spent years trying to unveil the mysteries that lie within such advanced knowledge systems and yet there
is not much that can be said to explain these complex systems. There

have also been countless conspiracy theories created to try to dismiss Indigenous ancestral science. Some of these conspiracy theories claim that extraterrestrial beings or giants helped Indigenous peoples build these structures. However, we know that these are just imaginary stories as there is no such evidence that points to the existence of both extraterrestrial beings and giants that could tower over a pyramid. They are just conspiracy theories because present-day Western knowledge cannot comprehend such advancements.

The elders of my communities always told me that our ancestors' advancements were far superior to that of the colonizers. As a result, during colonization, colonizers were afraid because they had never seen or witnessed such intellectual and technological advancements. This is why they created narratives that condemned Indigenous ancestral knowledge and practices. They often related it to worshiping the devil or evil spirits as colonizers were also motivated to spread Christianity. Yet today many scientists are trying to study ancestral Indigenous systems because they want to uncover the knowledge colonizers burned and eliminated throughout the Americas. I say "trying" because no matter how advanced scientific knowledge has gotten, it still cannot fully explain the creation of such complex systems like our pyramids.

One of these Western scientific fields that is still trying to unravel the complexities of these ancestral systems is physics. Physics is trying to explain, within Western knowledge and science, the building of the pyramids, yet it has not advanced its reasoning. Physics concepts have been applied, but these concepts cannot fully explain how pyramids and other complex structures were built. This highlights the limitations that Western science has as no matter how advanced it can get, it still cannot explain how many ancient Indigenous civilizations were able to build, generate, and complete such things with minimal modern tools. However, behind this ancestral Indigenous knowledge and their abilities also lies a deeply rooted grief, especially for those of us who are descendants from these amazing ancient civilizations.

Being a scientist and also an Indigenous Maya or Zapotec woman, I can reassure that there is a lot of grieving for the ancestral knowledge that was destroyed and burned by colonizers. It makes it difficult to see how scientists and archaeologists are still studying my ancestral histories and sacred sites as I know too well that this was a part of my identity that was destroyed during colonization. Any magnitude of loss hurts, and this is a loss that I know and feel when I visit my ancestral and sacred sites. It is one of the greatest losses my communities have faced as we are proud to be the descendants of the Indigenous peoples who built these pyramids and complex structures. Yet my inherited right as a descendant from these ancient communities that built great civilizations was stolen from me and my communities because of colonization.

Yes, Western science has provided society with great advancements that have also increased human life expectancy, but it has taken centuries to get where it is currently today. All I can imagine is how technologically advanced Indigenous civilizations and communities would have been today if their ancestral knowledge had not been burned and destroyed by colonizers. We get to keep some snippets of this history through artifacts archaeologists have been able to find, but many times these artifacts that belong to Indigenous communities are not rightfully returned to them. One of these examples is *codices*, which are written historical records that many Indigenous communities of Latin America sustained precolonization.

A lot of Indigenous ancestral knowledge from Mexico and Central America has been preserved in *codices,* or hieroglyphics that were the written records during precolonial times. However, as mentioned before, this was part of the ancestral history that colonizers burned and destroyed. Some of these codices were smuggled into Europe and have been displayed in their museums for years. Others were taken by colonizers as gifts for their monarchies and for other noble Europeans. For example, as Maarten Jansen, emeritus professor in Heritage of Indigenous Peoples at Leiden University, wrote in 1990, "The book on the origins of the Mixtee Lords and the genealogical history of Tilantongo, now known as the Codex Vindobonensis Mexicanus 1, for example, was in Europe by 1521, and thereafter experienced a turbulent and peripatetic history."[1] This peripatetic history points to the oppressive history behind many museums that have taken Indigenous ancestral artifacts or history and preserved it for visitors. Colonizers either burned and destroyed our history or decided to display it behind museum glass cases. However, when museums decide to keep Indigenous artifacts and ancestral histories instead of repatriating them to the Indigenous communities they rightfully belong to, it is also another form of ongoing colonization. What museums continue to do with many Indigenous artifacts and histories also continues to fracture Indigeneity because it limits the communities' right to revitalize some of this ancient history and knowledge that was lost.

Everyone gets to experience, witness, or see this displayed history and ancestral knowledge except the Indigenous peoples and communities the artifacts or histories belong to. This notion of displaying our Indigenous histories behind museum glass doors focuses on the colonial belief that Indigenous peoples and communities would become extinct. In Oaxaca many Indigenous communities have fought to have their codices returned to them, but their efforts continue to be ignored. Europe eventually returned some of the codices to Mexico, but the Mexican government gave them to some of the country's museums. Some of the codices that survived colonization belong to the Mixtec, Ñuu Savi, Indigenous community of Oaxaca. Their codices were written on deer hide and consisted of their writing style and format known today as logographics.[2] The codices were their written histories and many of them were transmitted orally or physically performed during their gatherings and ceremonies. Ms. Juana Silva, a Ñuu Savi woman, shared the importance of codices for her community and future generations. She is an artisan who is a member of the Ñaa Ñanga Tijaltepec Collective.

Can you explain why codices are important to the Ñuu Savi community?

They [codices] are important to our community because they are the written history of our pueblo, people, and grandparents. However, during colonization they were destroyed because they were associated with the devil. Given that many [colonizers] had not seen this type of writing, they mentioned that it was a form of worshiping the devil or evil spirits. They [colonizers] were interested in spreading Christianity to our pueblos, so they wanted to find something to justify their violent actions. They burned most of our codices and the only few that we know that survived were due to them being sent over to Europe. I remember a young man, who went to study in Mexico City, telling me that our ancestral codices were sent to Europe and were studied by European scholars. They wanted to study our history and mostly us, during colonization. However, to this day, our codices have not been returned to us. Last time I checked, I remember they were on display in some museum in Mexico City . . . I do not remember the exact museum they are displayed by, but they are not with us. I think it will be amazing and crucial for our youth to have our codices returned. This will allow us to reclaim and revitalize some of our ancestral knowledge that was burned and destroyed. Even if it's just learning again how to draw the figures or be able to study them. The number of youth who are attending college is growing little by little, so even being able to have them so they can study them will be very important for our community. I remember my grandmother telling me that a lot of the animals they used to embroider on our traditional garments were inspired by the codices. However, we have not been able to see them in person but only through photographs. They are published in some of the history books of Mexico that write about us in the past tense. We once had a young man who went off to study in Mexico City. He went to the museum and took photographs of them. When he came back to visit his mom, he brought back those photos and we all got together to see them. We have used some of those photographs to embroider new figures in our garments. We all mentioned that one day we wanted to go to Mexico City to see them in person. However, it is too expensive to travel, especially when you have children, so that trip might never happen.

Ms. Juana Silva shares something that is important to Indigenous communities: the transmission of Indigenous knowledge throughout the seven generations. She mentions that her grandmother's generation was trying to revitalize some of the knowledge from their codices by seeing photographs and embroidering them in their traditional regalia and garments. However, so much can be done through photographs as opposed to the physical item itself. If they had access to the physical codices themselves, the Mixtec community would be able to revitalize their ancient traditions and knowledge that were lost or fractured due to colonization. This is because within these codices not only were their histories recorded but also the environmental knowledge that helped these civilizations during their ancestral times thrive and take care of nature.

However, many Indigenous communities within the Mixtec, Maya, and Zapotec, among other pueblos, have lost part of this history due to colonization and continue to be denied the right to revitalize it due to museums choosing to keep Indigenous sacred items instead of repatriating them.[3] The history that made it through this era of colonization is now being kept behind museum glass doors, and this continues to be the harmful legacy that continues to lead to the theft of these sacred items. It is also important to note that some Indigenous communities faced more harsh conditions and atrocities committed by colonizers because they were one of the first Indigenous communities to experience colonization. This is not to generate a comparison or a discourse focusing on who got it worse but rather to acknowledge that for many Indigenous pueblos and communities, colonization was different because the colonization of the Americas began in their territories. The Indigenous communities that were first faced with colonization are those located in Latin America.

The Colonization of the Americas Began in Latin America

During the precolonial time periods of the 1500s, Europe had established trading relationships with Asia. Most of the trading was done through journeys via land voyages. However, some countries were hostile toward European nations, so the journeys on land were becoming inconvenient for many European nations. This is when they started focusing on ocean voyages. Their main goal was to create a new route so they could continue trading with Asia without having to go through countries that were hostile toward them. Unfortunately in 1492, when Christopher Columbus was provided with three ships by the Spanish monarchy, Columbus's ocean voyage did not necessarily lead him to Asia. It led him to the Americas, and it is important to mention the location that it led him to first was Latin America. He landed in Guanahani, which is now known as San Salvador, located in the Bahamas, and is an island that sustained and was inhabited by the Lucayan people.[4] It is important to mention that the Bahamas are located within the Caribbean islands, and they are considered part of Latin America. However, within North America, the 1492 colonization discourse fails to recognize that Columbus first set foot in what is considered Latin America today. His journey then transcended across the Caribbean, coming in contact with other Taíno, *Arawak*, communities. This means that Indigenous peoples and communities of the Caribbean, primarily the Arawak communities, were the first ones to experience colonization and the genocide and disease that this era brought to the Americas.

Given the differences that exist within Indigeneity from the United States and Canada due to tribal sovereignty and enrollment, it is important to ensure that Indigenous peoples from Latin America are not forgotten or ignored in this discourse as this is a major part of the Americas' history. Many Indigenous communities in Latin America did not sign treaties like most tribes in the United States or Canada did, because colonization impacted them first. There was no time to set colonizer-Indigenous relationships to ensure that their rights were somewhat protected. This is not to say that treaties today are respected by both governments in the United States and Canada, but with no treaties there is no legal framework that can be brought forward to the government when Indigenous rights in Latin America are dismissed or ignored. Despite Mexico having Indigenous rights written in its constitution, after its constitution was amended in 2001 (discussed in chapter 2), there are no set treaties that can be brought forward, and this is one of the main reasons why Indigenous rights violations are often dismissed in court cases within Mexico.

Bringing to the forefront the history of colonization's initial place as Latin America also allows for the discourse on colonization's impacts on Indigenous communities that resulted in the huge decline in their populations being highlighted as well. For example, upon Columbus's arrival, the Lucayan people, who are also an Indigenous group of the Taíno society, suffered severely. Their populations declined and many settler scholars have claimed these populations indeed went extinct.[5] However, it is important to note that using the word *extinct* toward an Indigenous community is oppressive given that people cannot go extinct, unless the entire species of humans does. As Tony Castanha, Lecturer of Indigenous and American Indian Studies in the Department of Ethnic Studies at Hawai'i Pacific University, has claimed,

And there are many of them [Tainos]. They populate the many bar-
rios of particularly the rural and mountain regions of Puerto Rico,
and coastal areas as well. Whole communities of Jibaro . . . people
have survived the Spanish and American colonization process and
continue to practice their cultural traditions today.[6]

Castanha is indeed right, as despite the genocide Indigenous peoples
experienced during the first contact with colonizers that came to the
Americas, especially in Latin America, their cultures never went extinct.
Their teachings have been preserved in their communities and one of
the main teachings and knowledge systems that has been preserved is
Indigenous stewardship. It is what has allowed many Indigenous communities to adapt to their new environments and continue to thrive in
their ancestral homelands despite all the impacts they faced and continue to face due to settler colonization. This highlights the importance of the nature-human relationship in regard to their resilience and
resistance.

Indigenous Stewardship

Indigenous knowledge has been able to survive colonization because it has
been passed down orally. While some of these teachings were in relation
to their living conditions in the past, Indigenous knowledge can adapt to
new climates, spaces, and environment. It is no surprise that due to their
resiliency and adaptive capacity, Indigenous peoples' teachings can serve
as solutions to the environmental degradation and crisis we are currently
facing in a changing climate. However, these Indigenous teachings lose
their holistic worldviews when applied to the Western linear way of thinking, as Western ideologies do not incorporate *culture* in their teachings or
beliefs. This is why it is important to incorporate Indigenous peoples and
their Indigenous teachings when trying to find solutions to mitigate the

effects of climate change into the environmental narrative. We cannot just solely identify Indigenous teachings and remove Indigenous peoples out of this narrative as co-option of Indigenous knowledge contributes to the oppressive narrative we currently have in the environmental discourse. These teachings do not and cannot be applied in the Western scientific paradigms or frameworks without incorporating Indigenous peoples as well.

This is why Indigenous communities are more careful of who they share their Indigenous teachings or knowledge with, especially sacred knowledge. There have been many examples of Indigenous sacred knowledge being co-opted and stolen to exploit capitalistic structures. Many Indigenous scholars and community members have experienced intellectual theft or co-option and it needs to be called out as it creates more harm against Indigenous peoples.

One of the fields that has been established within the environmental sciences is *permaculture*. Permaculture is co-opted global Indigenous knowledge that was introduced into the Western science world by Bill Mollison. He is now described as a pioneer who has observed nature's cycles and movements to re-create an ecosystem that provides food and other natural materials. However, Mollison learned how to observe the natural world through the Aboriginal Tasmanians' Palawa lens because he lived with them. Now permaculture is utilized as a strategy for ecological resource management and restoration. Permaculture is defined as a development of agricultural ecosystems intended to be self-sustainable and self-sufficient without dependence on factors to aid it.[7] Mollison is idolized for this discovery when in reality the Palawa and other Indigenous peoples have been living through this lens for centuries. This is why it is important to protect Indigenous knowledge of soils and lands as it can either be ignored or co-opted to gain recognition and wealth.

Indigenous peoples have sustained Indigenous stewardship through their immense knowledge of agroecology; however, in this field of agroecology Indigenous peoples are often ignored and dismissed.[8] Permaculture is still a growing field, and classes to get certified to become a permaculture designer or teacher range from $1,000 to $5,000.[9] None of the funds are going to the Palawa people or other Indigenous communities, and while they are mentioned briefly in the description of the founding of permaculture, they are not given credit for their way of life. Since permaculture is Indigenous co-opted knowledge and is seen as effective in conservation efforts, it is important to indigenize conservation so that we can create holistic solutions in a changing climate. We must do so by uplifting Indigenous voices and not stealing or co-opting their knowledge.

I received a certification through an Indigenous program for permaculture that was led by a nonprofit in Seattle. Upon immersing myself in this course, I started questioning the practice of permaculture given that it was indeed co-opted Indigenous knowledge by a white man who continues to be credited as the founder of this field when in reality he just stole this knowledge about sustainable ecosystems from Indigenous peoples. I asked the instructor if we were going to uncover the colonialism behind this field and the instructor replied, "We have to give him credit for bringing this to us, and in a way he decolonized permaculture." I was shocked at the instructor's response because, to me, decolonizing does not mean co-opting or stealing Indigenous knowledge to make a huge business of certifications, classes, and training. Indigenous *epistemologies* need to be respected and should not be stolen to make settlers money or generate recognition for them.

Indigenous knowledge systems cannot be defined by one sole definition as it consists of all Indigenous knowledges—locally, regionally, and globally. While some Indigenous knowledges can share similarities with one another, they differ based on the geographic location of each tribe, nation, or community. This is due to the reality that Indigenous knowledge is place based and not socially acquired through Western systems (i.e., education). Indigenous knowledge is acquired through cultural upbringings and Indigenous ways of teaching (i.e., storytelling). It is not entirely written in textbooks or expressed in research as some components of Indigenous epistemologies are sacred and are only kept between family and tribal kinships. Indigenous knowledge systems include both traditional ecological knowledge and Indigenous knowledge.

It is important to note that tribes and Indigenous communities have the autonomy to define their way of thinking, knowing, and being in the world with the term or concept of their choosing. As a result, we will use terms that define Indigenous ways of knowing interchangeably in this book. Some of these terms include *tribal science Indigenous knowledge, traditional ecological knowledge, Indigenous ethnoscience,* and other terms each community member utilizes to describe and label their knowledge.

Something I have witnessed as an Indigenous scientist is how Indigenous terminology is being co-opted to validate settler colonialism in the environmental sciences. For example, when I questioned the colonial legacy behind permaculture, the instructor used the word "decolonizing" to validate what Mollison did with the Indigenous knowledge the community shared with him. Personally, there is nothing decolonial in stealing and co-opting Indigenous knowledge, especially as Indigenous peoples continue to face the impacts of settler colonialism. Decolonizing along with other Indigenous words have been co-opted despite the immense Indigenous work that has been put out there like "Decolonizing Is Not a Metaphor" by Eve Tuck and Wayne Yang.[10]

Throughout my academic journey in graduate school, I was asked to cite my Indigenous knowledge that I would share in my writing, which placed this layer of settler colonialism that invalidates Indigenous knowledge. Yet, when settler scholars (white scholars or non-Indigenous scholars) write about Indigenous knowledge or communities, they are not asked to cite as intensively as I have been asked to do in my entire life. This adds to the constant advocacy Indigenous scholars have to do to ensure that their knowledge and ideas are not stolen or co-opted in white-dominated spaces like the environmental sciences. For example, when I was a graduate student at the University of Washington College of the

Environment, I advocated tremendously to be able to teach the first environmental justice course from an Indigenous lens.[11]

Now there are white/settler professors who are teaching environmental justice courses or Indigenous science courses who should not have such positions when there are countless Indigenous scholars who are qualified based on their lived experiences. However, academia teaches us that there is this need to publish in order to gain qualifications, yet these publishing opportunities are taken from us when settler scholars steal our knowledge or ideas and publish this work themselves. Indigenous scholarship has become co-opted, and as a result, a lot of articles and publications about Indigenous peoples and knowledge are written about them and not by them.

This is one of the reasons why I was motivated to serve as a guest editor for a special edition of the journal *Human Biology* about Indigenous science and ecology. I asked my coadvisor Dr. Mike Spencer to serve as my guest coeditor, and one of the things we decided was that the first author had to be an Indigenous scholar or community member. Like I mentioned before, many settler and white scholars think that they are being inclusive of the Indigenous communities they are working with by making one of their community members a coauthor when they publish data they collected from the Indigenous community. However, it is rare for them to support having the Indigenous scholar or community member write the article instead of serving as a coauthor. We did receive publications with a white or non-Indigenous first author and we had to request that the authors change this because the issue was *Indigenous Science & Ecology,* not *Western Science & Ecology.*

It is important for scientists and academics who are working with Indigenous communities to reflect on their positionality as this can cause more harm. In academia, scholars and scientists are taught that productivity and publications are important, because these are weighting factors in higher academic careers. However, many see publications on Indigenous knowledge as an opportunity to advance their careers without reflecting on the harm they are causing when they as a non-Indigenous or settler scholar publish Indigenous scholarship. For me personally, Indigenous scholarship published by a non-Indigenous or settler scholar is a form of colonization because Indigenous knowledge does not rightfully belong to non-Indigenous or settler scholars. It belongs to the Indigenous community and oftentimes, in publications from peer-review journals to books, Indigenous peoples are rarely given opportunities to publish because the metric used in these publications requires an advanced vocabulary or terminology that is rooted in Western academia. Academic opportunities are rarely offered to Indigenous communities across the Americas, and as a result, the number of Indigenous peoples who can have such advanced terminology and curriculum is limited.

Daniela shares her pueblos experiences of how non-Indigenous peoples have harmed her community with the co-option and theft of Indigenous knowledge:

> *How has Indigenous knowledge been co-opted or stolen? Any examples from your pueblo?*
>
> Indigenous knowledge has been stolen from my pueblo by non-Indigenous teachers refusing to pass down the knowledge and educate my parents' generation on how to take care of the land. There was a tremendous amount of shaming for our Indigenous culture during my parents' generation and knowledge about the land was included in this shaming. Because of this there was a lot of deforestation allowed by outsiders.
>
> *How does this cause harm to your pueblo and other Indigenous peoples?*
>
> The harm that this has brought comes in robbing my pueblo of knowledge passed down from our ancestors that cared

about the land. There is a whole generation that has missed out on this knowledge and now can't pass it down to their own children so essentially it's two generations that don't have this knowledge. Indigenous knowledge that is passed down by word of mouth and could be lost forever when the older generation is gone.

Daniela shares how her pueblo has been impacted by Indigenous knowledge theft and co-option. She points to something important that many of those in our Indigenous elders' generation had to face: being shamed for their Indigenous knowledge. When thinking about this perspective, many Indigenous peoples continue to be shamed for their knowledge because it is not deemed as valid or as intricate as Western science. When I think about how many white and settler professors asked me for citations for my knowledge, there was always a ridiculousness in their voices and way of saying this statement to me. They never asked other students to cite any of their lived experiences, but then again, those students were white, as were a majority of the cohorts in the master's program I had completed. Their stories about fishing with their parents or traveling in other countries outside the United States never demanded citations, while me sharing my stories of how my communities interacted with their environment did. This is the shame that continues to be placed on Indigenous knowledge, as though it is something to truly be ashamed of.

However, constantly being shamed for the knowledge one possesses leads to internal conflicts as one starts to think that indeed our knowledge is not worthy. This has happened across Indigenous generations, and as Daniela pointed out, it leads to knowledge and cultural loss across the generations. This conditions Indigenous communities to have to rely on settlers or non-Indigenous scholars to reteach them that Indigenous knowledge, yet it is not taught to the Indigenous communities by scholars, teachers, or researchers who have spent their lifetime extracting Indigenous knowledge from the communities. This happens a lot across Indigenous languages as settlers and non-Indigenous linguists go to our communities because they have the resources to document and study native languages. Yet, many Indigenous youth that were displaced do not have access to learning their native language. We will dive into Indigenous languages in chapter 5.

Many Indigenous peoples have a rich history of ecological management yet have been subjugated and exploited in the name of conservation. Given that the ancestors of Indigenous communities of the Americas had thrived and interacted with their local environments since time immemorial, their knowledge about their local environments and natural resources was superior to that of Western science. Western science is linear and binary as it seeks to find an answer to a question. It follows the scientific method that starts with asking a question and then formulating a hypothesis, an idea to try to explain or answer that question. Indigenous peoples knew already which plants they could consume, which plants were lethal, and how to hunt to ensure they had sustenance. This knowledge was developed over time and allowed them to thrive within their local environments and advance their technology.

Their deep understanding of their local environments allowed them to steward a high biodiversity that attracted colonizers, who were in search of wealth. Most colonial countries in Eastern Europe had already destroyed their environments and landscapes due to their extractive practices.[12] For example, in England, during the middle Holocene time period, over five thousand to seven thousand years ago, the agricultural societies they built had resulted in massive deforestation. Their heavy reliance on wood products for fuel and construction also contributed to this deforestation. This is due to the new way of life Europeans had adapted that shifted their societies from being foragers to farmers. This shifted their lifestyle, and they started to primarily focus on food production through agricultural practices such as domesticating animals for consumption.[13] The desire of the European monarchies to find new lands so that they could use the lands' natural resources for their way of life is what motivated them to invest in ocean voyages that resulted in the colonization of the Americas.

Kincentric Ecology

Our relationship with our environments differs and it is important to mention this as we are not monolithic groups and our relationships are place based. Some of these environmental relationships as previously discussed are inherited based on where our ancestral lands are located or are those we have cultivated through the diaspora. This human-nature relationship has been described as *kincentric ecology,* and this term has been further developed by Indigenous scholar Dr. Enrique Salmon.[14] This term tries to explain the human relationship Indigenous peoples have with their environments through the notion that we are not separate from nature but rather an integral component. Kincentric ecology describes how Indigenous peoples view their natural resources and surroundings as part of their kin, relatives, and communities. He describes how his Rarámuri community embodies kincentric ecology through *iwígara,* their way of life and worldviews that make nature an essential component of their existence.

> The Rarámuri view themselves as an integral part of the life and place within which they live. There is among the Rarámuri a concept called iwígara, which encompasses many ideas and ways of thinking unique to the place with which the Rarámuri live. Rituals and ceremonies, the language, and, therefore, Rarámuri thought are influenced by the lands, animals, and winds with which they live. Iwígara is the total interconnectedness and integration of all life in the Sierra Madres, physical and spiritual.[15]

For my Zapotec community, part of our life, *guenda nabani,* is to protect and embody nature as our relatives and part of our identities. This kinship or relationship in my Zapotec community is also manifested in our embroideries, which are a major part of our traditional regalia. Many of our embroideries are flowers and nature based, inspired by our landscapes and local environments. My grandmother always told me,

When you wear a huipil you are wearing a part of nature. This is how you can take your beautiful ancestral homes of Oaxaca everywhere you go.

PHOTO 4.1: My cousin Dalia with her son Jonathan. Photographer: Eduardo Salinas Villatoro

Through our huipiles we weave together our landscapes and plant relatives. For the Zapotec community, flowers are important symbols because Oaxaca is one of the regions that has a high flora biodiversity.[16] Even our head crowns that are part of our full regalia are made up of flowers. This shows the immense importance and strong relationships we have with our environments. Our embroiders, which I will discuss more in depth in chapter 6, weave years of resistance and cultural understanding of our environments despite colonization. They serve as this symbolism that ties everything back to relationships with nature, kincentric ecology, as it is now acknowledged and known. Kincentric ecology and the kinships we have with nature are the main reasons why we continue to advocate so that as Indigenous peoples, we are allowed to manage our own natural resources.

Without managing our environments, capitalism and extractive natural resource methods continue to harm our environments. In the community of Santiago Xiacuí, part of the Zapotec nation, over three hundred *comuneros,* Indigenous communal environmental stewards, are applying community-based forest management (CBFM) to ensure that they not only protect their forests from illegal logging and deforestation but also protect their livelihoods.[17] The forest of over eight thousand meters is managed communally by the Zapotec pueblo of this region.[18] CBFM allows conservationists and environmental scientists to learn a lot about the misguidance they have followed for generations, which removes human presence or impacts from the environment.[19] This continues to separate that reality that, for Indigenous communities, part of stewarding their environments and caretaking of the land provided them with everything they needed to survive.

CBFM also allows Indigenous communities to practice intergenerational learning as elders teach adults about ancestral forest management, and then the adults teach the youth while both work side by side in the forests. This ensures that this traditional knowledge of stewarding the forests is not lost. This also demonstrates the importance of allowing

Indigenous peoples to manage their natural resources, as it also allows for Indigenous knowledge to be maintained within the Indigenous community for the future generations. For the Zapotec community, *gulabere'*, trees, are important to their Indigeneity as trees have been taken care of and utilized as part of ceremonies, cultural traditions, and medicinal plants since time immemorial.

The Zapotec pueblo of Santiago Xiacuí is also able to provide educational scholarships to Indigenous youth who decide to attend school and obtain an education from the funds generated from the forest products that this sustainable forest provides them. This is not only how they are they able to manage their own natural resources but also how they increase the job and educational opportunities offered to the youth. It also embodies the bilateral leadership the Zapotec communities have been able to sustain for years and the communal harvests and, in this case, forests that are important to many Indigenous pueblos of Oaxaca.

Contribution of Indigenous Ancestral Practices to the Wealth of Biodiversity Present Precolonization

We have all read a version of the statement "While the world's 370 million Indigenous peoples make up less than five percent of the total human population, they manage or hold tenure over 25 percent of the world's land surface and support about 80 percent of the global biodiversity."[20] This statement demonstrates the importance of Indigenous knowledge, kincentric ecology, and ways of being. Colonization decreased the access to land Indigenous communities had precolonization because land became privatized. Land also became property that continues to be treated as such. The extractive methods settlers introduced into the Americas continues to also lead to environmental degradation. Many read this statement but do not fully understand what this statement is actually mentioning beneath the surface. This statement is pointing out the fact that despite receiving less land under settler colonialism, Indigenous populations declining, and Indigenous peoples only being granted certain rights, they are still maintaining their stewardship of the lands they occupy. Their stewardship is so intertwined with their environment that they are able to sustain 80 percent of the world's biodiversity despite only living within 25 percent of the land that is available globally. Yet Indigenous peoples have to constantly continue to fight against their right to preserve and manage their environments because land is always being stolen the same way Indigenous knowledge is also being stolen. It is their Indigenous knowledge and the way they have created communal management systems like many Indigenous pueblos in Oaxaca, from stewarding their milpas to their forests, that they are able to ensure the biodiversity that embodies 80 percent of the entire biodiversity that is still available today. This shows how informing their Indigenous knowledge is and how all the global Indigenous knowledge can create a tapestry of systems that are maintaining the overall health of our planet.

It is important to reemphasize that Indigenous knowledge and teachings are described as *place based*. This means that every Indigenous tribe, pueblo, or community has their own unique ways of thinking and managing their landscapes. Place based for Indigenous peoples goes more in depth than just an enclosed natural place. It broadens to the landscape, and this more holistic lens is embedded among Indigenous knowledge systems. To contextualize this, I asked my father about his journey in learning how to become a successful fisherman who was self-taught.

> *Dad, you have told me on various occasions that you taught yourself how to fish. Can you explain how you were able to learn?*
>
> It is interesting to see how teaching in many cultures outside of ours is very linear. Where you have someone teaching you directly, like a teacher and you are just learning. I never understood how one was able to learn within this type of structure. I also was never given the opportunity to go to school, as my father had passed away at a young age, so I had to work to

support my mother and younger siblings, so I do not have that experience of how effective this model is. However, for me, I was always told by my elders when I would go ask them how one should fish to observe everything within this environment where the streams, rivers, or oceans were located. To not just stare at the water where the fishes were, as this was not going to teach me how to truly catch fish. Every day I was instructed to look at something and observe how it interacted with the waters. I remember being told to observe the trees and see how the leaves fall. This ultimately taught me about the wind movement. I learned when the wind was moving south based on how the leaves were falling to the water. The next day I was instructed to observe inanimate objects like rocks and pebbles located in the stream, without forgetting to observe how all of this interacted with the water. The elders that were teaching me could no longer walk or move physically, so everything they were teaching me was verbal. I think being able to observe everything that essentially was a part of the stream, river, or body of water where I wanted to fish allowed me to learn at a young age that our ways of learning and knowing looked at the entire landscape, and not just at segments of nature. It is hard to explain how we [Indigenous people] view our environments differently than others, because there are no words in Spanish or English to fully describe it.

Observations helped you understand how everything was inter-connected to the water and ultimately the fish. What other things were you instructed to do so that you could learn how to, for instance, create tools and materials to help you catch the fish?

I was always instructed to use natural products to create the fishing rods or nets I would use. Sometimes these tools are lost in the water, and you do not want to end up leaving plastic or something dangerous in the water. For our community, the river that passed through our *cantón* was also our form of bathing, so it was important we protected each other as well from getting hurt. I think that we always tend to learn how to see everything linearly, from point A to B, etc. This does not allow us to understand our overall environment. It is easier to just focus on the water and the fish if your goal is to learn how to fish, but this is, like I mentioned, the way non-Indigenous peoples learn. Where they are taught to just look at two points max . . . and ignore everything else. My elders taught me that everything is a part of the ecosystem, more like a landscape as opposed to just a river, a fish, a tree, a rock . . . everything is interconnected in this web of life, and landscape.

My father's responses to my questions about how he was taught to look at the environment only show a snippet of everything he learned. He mentions how hard it is to describe our ways of being and looking at the world through Western terminology. Yes, an Indigenous scholar presented kincentric ecology, but there is still much sacred knowledge and many ways of being that cannot be embodied by this term. Indigenous peoples tend to look at landscapes holistically as opposed to looking at them through systems that are somehow connected. I once had an argument with a physicist after I mentioned that Indigenous knowledge is holistic because of the way we view our environment. His argument was that this was *systems thinking* and that physics did that as well. However, it is important to note that non-Indigenous and settler scholars or scientists should stop attempting to describe Indigenous knowledge or ways of being through Western knowledge frameworks and terminology as this fails to explain the complexities of our cultures. It is also oppressive to be told by settlers and non-Indigenous peoples how we actually think when in reality, we know best about our way of life and lived experiences.

Systems Thinking Is Not
Indigenous Holistic Thinking

Indigenous holistic thinking is not systems thinking because many Indigenous peoples and communities do not separate their world or landscapes into "systems" in the first place. I think that this is why the physicist's argument over how holistic thinking is system thinking confused me, because as an Indigenous person I was never taught to think of everything separately. This is why pursuing Western education and higher academia made it difficult for me to truly comprehend the importance of this way of thinking as it compartmentalizes everything. Western knowledge has separated everything into either systems or boxes, which is why there is an array of disciplines within the environmental sciences and academia. If someone were interested in learning more about fish health or biology, they would pursue fisheries. If someone were more interested in learning more about the oceans, they would pursue oceanography or marine sciences. While these disciplines, fisheries and oceanography, are two different systems, through "systems thinking" they can be connected. However, for Indigenous peoples, separating our world into systems is why many environmental or climate solutions are not effective and continue to fail to address the root of the problem. This is because when everything is separated into systems or boxes, more harm can be done with the solutions that are thought of. For example in 2018, Seattle passed a ban on plastic straws. This meant that restaurants or any places could no longer provide plastic straws and they had to provide either paper or other recyclable straws. However, with this ban on plastic straws, disabled individuals no longer had access to an essential tool and material they needed. Alice Wong wrote the following testimonial on why plastic straws are essential to disabled individuals in the article "The Rise and Fall of the Plastic Straw":

> A plastic straw is an access tool I use for nutrition as a person with a neuromuscular disability. When sipping a latte at my favorite cafe, I use

a plastic straw because I am unable to lift a drink to my mouth and it is safe for hot liquids. Plastic straws are now seen as harmful and out-moded by environmentalists who are in favor of "safer" products (e.g., compostable, biodegradable plastics made of polylactic acid, silicone).[21]

Wong's testimony reveals the nuances and intersectionalities that are often dismissed within this systems thinking. The systems that came into play in this decision aimed to reduce the impacts of plastic straws, yet due to not linking the system of disability justice, they ended up causing more harm. Therefore, eliminating plastic straws is not an equitable decision given that it further harms a marginalized community. This is why environmental solutions are not as inclusive because we all know who these systems are primarily governed by. They are governed by those who hold on to power and privilege provided within the systems under settler colonialism that the United States operates under. Given this, it is also important to mention that as long as every system is influenced by and operating under settler colonialism, it will never be equitable or just toward Indigenous peoples. This is why many Indigenous peoples do not think within this system's framework and everything is rather holistic. If we were to try to integrate systems thinking into Indigenous ways of knowing, I will say that we think within one system that encompasses everything. As my father recounted in his interview, it is hard to fully explain how we as Indigenous peoples think, but we know that everything is interconnected with our environments. This is why our worldviews as Indigenous peoples are distinct from Western worldviews.

Everything is interconnected, even during our environmental and climate justice movements. We do not just advocate for our rights and natural resources, as it should be if we were applying this systems thinking into our ways of knowing. We also advocate for language, gender, spirituality, and everything else that is integral to our identity as Indigenous peoples. Everything is interconnected ultimately to our environment through our cultural values and ways of knowing.

5

Ecowars:
Seeking Environmental Justice

*W*hat is the hardest thing you have to face every day as a woman of *color in your field?*

I was once asked this question and it really made me reflect on my positionality as an Indigenous woman in the environmental sciences. Yes, I have to face and constantly continue to learn how to navigate racism and settler colonialism every day because the environmental sciences continue to be deeply rooted in settler colonialism by only uplifting Western ways of knowing and invalidating Indigenous knowledge. Thus there is an inherited racism deeply embedded in the environmental sciences that also contributes to "othering" Indigenous and non-Western cultures. I recall when I was once called an "Indigenous cultures expert" by a white person. This made me reflect on why I had been positioned in such a way that was othering me and making me an expert on my embodied identity. It also made me question whether other identities were also othered, because no one ever labels themselves an expert on their embodied identities.

This also highlights how Indigenous cultures and communities are considered an "area of expertise" within the environmental sciences. As I previously mentioned, gaining experience with working or facilitating a research project that includes Indigenous peoples and communities has advanced many careers. However, those with lived experiences as Indigenous peoples continue to be left out from the environmental discourse. In reality, lived experiences should outweigh academic or work experiences as no one will truly ever understand what it is to be Indigenous, unless you navigate your world and life as an Indigenous person.

Even within this notion of being an "Indigenous culture expert," it is important to mention that Indigenous communities and peoples are not monolithic as our cultures and communities embody different intersectionalities. We are often considered a monolithic group throughout the Americas, without those labeling us understanding how coastal Indigenous peoples are different from more inland Indigenous communities. Even within those tribes and pueblos, everyone is different as we all have different intersectionalities that sometimes make us disagree with one another. There is no correct way to be Indigenous and this is the beauty that is embedded in our cultures and Indigeneities.

In Western ways of knowing, everyone also aims to reach a level of expertise in certain topics or discourses. However, I was taught by my elders that expertise is an ongoing process as we always have something to learn and, in some cases, unlearn and relearn because the way we have been taught has not always been equitable, inclusive, or understanding of non-Western ways of knowing. I choose to never consider someone an expert because no one can ever learn everything there is to learn in their lifetime, which is one of the Indigenous principles that has been sustained among my communities. The elders in my community have always reminded me that this was also one of the main reasons why my ancestors were so advanced in their ways of knowing and continued to grow their in-depth understanding of our world. They continued to formulate questions and find answers to the mysteries that lie within our universe and never thought or assumed they had learned everything. Expertise was something they also did not seek, but rather they sought to enhance their relationships with our environments and enrich their holistic way of thinking. This continues to contradict the scientific method as its linearity teaches scientists that once they find an answer or a couple of answers to their question, they have become an expert in that subject matter. It is difficult to know, as an Indigenous woman, that Indigenous cultures continue to be seen as a "subject matter" that many environmental scientists want to gain expertise in, because Indigenous cultures are not for consumption in any way, shape, or capacity, and thus no one can ever become an "Indigenous cultures expert." I think this is important for environmental scientists to keep in mind as we transition toward a just world that not only centers environmental, food, or climate justice but also racial justice.

This is just an example of how settler colonialism continues to be deeply ingrained in the environmental sciences by teaching people that they can indeed seek such expertise despite it being at the expense of communities who continue to be harmed by settler colonialism. So yes, racism, settler colonialism, and being othered have always been present in my experiences as an Indigenous woman. However, the hardest thing I have to face every day is knowing that for the work I am doing and leading to protect our lands, Mother Earth, and advocating for Indigenous rights in the United States, my relatives and community members who are land and rights defenders back on my ancestral homelands in Latin America are targeted every day. Land and rights defenders are Indigenous peoples who advocate for the protection of natural resources and Indigenous rights, sometimes going against large multimillion-dollar corporations and settler government structures. Being advocates and defenders of our Indigenous rights and lands in Latin America means that many of our Indigenous leaders and advocates face extreme violence, persecution, and death.

For Indigenous peoples, facing the loss of a loved one, community members, and relatives due to their advocacy for land and Indigenous rights is something hard to face or ever experience. Every year, since I became a young adult and started understanding more complex concepts like murder, I started realizing that I was losing many important people in my life because of their advocacy. They were being assassinated because their advocacy was threatening someone with power and privilege, including multimillion-dollar corporations. Yet there is little acknowledgment of what happens to our Indigenous leaders who advocate for our land and Indigenous rights in Latin America. This is why it is hard for me to grapple with the reality that Indigenous cultures and people are deemed as areas of expertise within the environmental sciences when our Indigenous leaders and advocates continue to face alarming rates of violence for their lived and embodied experiences. Personally, knowing that my life is in constant threat when I visit my ancestral homelands makes it hard for me to wrap my mind around this desire for non-Indigenous and settler scientists to want to study us.

This is also a minimal glimpse of what many Indigenous leaders and advocates have to face daily. This was one of the reasons why I took on writing this book starting in 2020. I was trying to navigate life during the pandemic, natural disasters, and the extreme loss I was enduring. However, what kept me motivated to start and complete this book was knowing that too often our stories as Indigenous peoples are told for us and not by us. I think it is crucial that within the environmental sciences, we call out this settler way of thinking that continues to make Indigenous cultures and communities areas of expertise while invalidating and dismissing the lived experiences of Indigenous peoples.

Given that 80 percent of the world's biodiversity is sustained by Indigenous peoples environmental scientists need to focus on protecting Indigenous land and rights defenders' lives. In 2019, twenty-eight Indigenous rights defenders and leaders were murdered in Latin America.[1] Since then, the numbers have surged and Latin America has been considered the most dangerous place for Indigenous rights and land defenders.[2] This is something very difficult to grapple with as an Indigenous woman from Latin America. However, most of these deaths and murders go unsolved and justice is rarely served for the victims, their families, and their communities. The governments fail to provide Indigenous communities with resources to help them find the culprits behind their murders. It is no surprise, given that they fail to provide any protection to land defenders and Indigenous rights advocates to prevent this violence and decrease the murders in the first place.

These ongoing acts of violence and murder against our land defenders and Indigenous rights leaders is what I refer to as the ecowars. I understand that *war* should sometimes not be used to describe conflicts, especially as someone whose father was forced to fight in a war as a child, but it is very parallel in the sense that Indigenous leaders are targeted and murdered, sometimes with the same tactics the *escuadrones de la muerte* used during the Central American civil war in El Salvador, including targeting Indigenous leaders in broad daylight. This is a common tactic used because those in power want to use *fear* to decrease the advocacy that Indigenous community leaders take for their land and rights. It is time that this violence faced by Indigenous rights and land defenders in Latin America is acknowledged and that something is done about it.

Anything that impacts Indigenous peoples ultimately impacts our global biodiversity. Despite settler colonization, racism, and other injustice, Indigenous communities continue to steward a majority part of the world's biodiversity. When they are threatened, the world's biodiversity is also threatened. These ecowars have been occurring since settlers first stepped foot on our lands, as it was also important for our ancestors to protect our communities and environments. Unfortunately, due to all the beliefs and value systems colonialism introduced into the Americas, Indigenous rights and land protectors are a threat to every settler government. Given that capitalism and extractive resource methods continue to govern many countries' economies, natural resources are highly valued and sought after. Anyone who gets in the way of the greedy corporations that sustain these settler colonial values and beliefs is persecuted. Advocating for Indigenous rights ultimately means that Indigenous peoples will be given some power and autonomy to determine what governments can do with their land and natural resources. This is why Indigenous rights advocates also threaten these governmental systems that continue to profit off Indigenous lands. Since Indigenous peoples' rights are often-times violated, dismissed, or ignored, it is not in the best interest of these large corporations or any entities with power to grant Indigenous peoples any type of rights. The number of Indigenous land defenders and leaders murdered continues to increase, and this violence, coupled with the 2020–2021 pandemic, impacted Indigenous communities severely and increased their vulnerability during these years.

Indigenous Rights

Indigenous rights aim to protect Indigenous cultural values and systems that postcolonialism continues to threaten. This includes everything that is interconnected to our environment, culture, health, economy, politics, and other systems. This is why our Indigenous leaders focus not only on our natural resources but on other assets of our existence as Indigenous peoples as well. As I discussed in chapter 4, Indigenous worldviews are holistic and everything is ultimately connected to our environments. From our languages to gender rights, everything is a component of a holistic framework that interconnects back to our environments, landscapes, and natural resources.

Indigenous peoples do not place ideologies or concepts in boxes like Western knowledge and systems tend to do. Everything is connected, and this is why Indigenous and land rights range from gender rights to language preservation. This is also because, for Indigenous peoples, everything is impacted or threatened under settler colonialism. Colonization had one major goal and that was to steal land, extract natural resources (because Europe had limited resources during this time), and eradicate Indigenous communities. Colonialism thus introduced genocide, assimilation, and extractive methodologies into the Americas. This continues to exist and be implemented in modern times. Modernization has allowed people who have power and privilege to live a comfortable life, but modernization continues to threaten our way of life as Indigenous peoples because new technology is heavily dependent on the extraction or destruction of our natural resources. Without our natural resources, we no longer are living within the realm that carries our creation stories and sacred spirituality. With our sacred sites being desecrated to become national parks to our forests facing high deforestation to raise cattle for consumption, we are living in a time period when Indigenous rights should be lifted and placed in the center of many environmental conversations.

As Indigenous peoples, we know we will take action if no one else does. This is why our Indigenous leaders are not just verbal advocates but also continue to lead ground efforts to bring our Indigenous rights to the forefront. However, this continues to threaten those who have become comfortable; they target our Indigenous leaders as a result. When our Indigenous leaders advocate for anything from our languages to our gender rights, they become the enemy of those who are granted power and privilege within these harmful systems.

Oftentimes settlers and non-Indigenous peoples associate Indigenous rights with just natural resources, but Indigenous rights embody everything that is an integral component of our Indigeneity. Our native language is what sustains our ancestral knowledge of our environment and history. Our gender rights not only grant Indigenous women but also *muxes* their inherited rights to practice their traditions that are interconnected to our environment. *Muxe* is a third gender in our Zapotec culture, outside of the binary gender identities identified in Western culture, that refers to a male assigned at birth who also embodies the female spirit and duality. For Indigenous people, it is our right as Indigenous women and muxes to practice our spirituality to maintain our knowledge of medicinal plants that are native to our region. Everything is ultimately connected to our environments for Indigenous peoples of Latin America and around the world. This is why the Indigenous rights our Indigenous communities advocate for focus not only on our natural resources but also on our entire being and existence as humans. In our environmental justice movements, we advocate for the protection of our lands, our health, language, and other cultural values that Indigenous peoples hold important and crucial in their communities. There is no native word for "conservation" in my Zapotec language, but there are words that translate to "healing" and "protecting," and this is our role as Indigenous peoples that must be sustained with our environments. We are not just meant to heal or protect our lands but also heal and protect ourselves, our communities, and our people so that we can continue to survive settler colonialism.

The History of Environmental Justice in the Americas

In order to discuss environmental justice and how it is manifested among Indigenous communities, it is important to mention that the term *environmental justice* was coined within academia by Dr. Robert Bullard, who is now known as the "father of environmental justice."[3] He coined this word from the movement a rural and predominantly Black community in Warren County, North Carolina, led during the 1980s. The state of North Carolina did not have a dumping site to get rid of their toxins; therefore in 1982, the government decided to create a 142-acre dumping site in Warren County.[4] In this landfill more than eighty thousand tons of contaminated soil with thirty-one thousand gallons of polychlorinated biphenyls (PCBs) were dumped without ever consulting the community or asking for their consent. PCBs severely harm our ecological and human health, which is the main reason why their production was banned in the United States in the 1970s and in Europe in the 1980s.[5] PCBs affect our neurological, immune, and reproductive systems and can also lead to many types of cancers. So why was it easy for government officials to dump thirty-one thousand gallons of PCBs in Warren County? Given that Warren County was a predominantly Black community, it was easy for officials to make this environmental decision.

We as a nation and across the Americas continue to witness how Black and Indigenous communities are treated like landfills without the decision makers taking into account how it is going to impact the community's well-being and livelihoods. We have seen this manifested across several cases that have occurred in the Americas, such as the Dakota Access Pipeline.[6] Warren County's residents were outraged that their backyards were turned into a toxic dumping site. Their main concerns were the health effects their children and future generations would experience due to residing in close proximity to this dump site, and as an act of resistance and resilience, they held protests and marches to advocate for their environmental rights. During this time period, the United States had also witnessed the civil rights movement. The civil rights movement also motivated Indigenous rights movements across the Americas, which is why it is important to acknowledge the role the civil rights movement played in the environmental justice movement across the United States. This is how powerful this movement was and continues to be within American history.

Given this brief history of environmental justice, it is important to note that while it was a community-led movement and community advocacy, *environmental justice* is an academic term. For my Indigenous communities, environmental justice is not a term they use to describe their movements for Indigenous and land rights. When I taught courses around environmental justice, many community members who came to speak to my classes mentioned how *environmental justice* was not how they described their movements given that many students would ask them questions that included the term environmental justice. Many Indigenous communities do not see their advocacy or resistance movements as anything else but a means of survival or their inherited right. This is not unexpected given that this is an academic term that was popularized during the 2010s.

Similar to the Black community of Warren County, Indigenous communities continue to lead environmental movements to protect themselves

133

and future generations. One of the Indigenous principles roots itself in protecting our Mother Earth for the seven generations. It motivates elders to continue leading and advocating for Indigenous rights. Given the advocacy behind the Black and Indigenous communities of the Americas for environmental justice, it is important to mention that despite this being an academic term, it is deeply embedded in the community. However, environmental justice has been a term that has been co-opted by many because sometimes their use of the term does not lead to any action, implementation, or community-based work that centers the community seeking environmental justice. It is interesting to see how developing a persona as an environmental justice advocate or researcher is another way that Indigenous and communities of color–led movements have been exploited by many for career advancements. However, we must not forget that behind environmental justice, there are communities' livelihoods and well-being at stake. Communities of color do not choose to do this work but rather have to do this work as a means of survival for them and their future generations.

In the Americas, I believe that within the environmental justice discourse, Indigenous peoples from these lands should be placed front and center. Oftentimes the narrative that Indigenous peoples must fight and advocate for themselves to get a seat at the table is spread by politicians and other political actors. However, sometimes the best thing for Indigenous communities to do is create their own table so that they can place their cultural values at the front and center of it.

The way environmental justice is addressed present-day is still grounded in Western ways of thinking, and not just because it is an academic term but also because of how it has been implemented within policies. In the United States, for instance, environmental justice is addressed through three legal processes: distributive, procedural, and process justice. Distributive justice advocates for the equal distribution of goods and impacts, procedural justice incorporates fairness to resolve disputes over the allocation of resources, and recognition justice acknowledges the inequities and who it affects.[7] However, these three frameworks continue to not include cultural values that are important to Indigenous communities. Thus settler government policies do not fit the Indigenous experiences of the Americas as they are deeply grounded in this settler framework and theory of environmental justice that never incorporate cultural norms, values, and principles.[8]

Indigenous rights and advocacy go beyond just focusing on fairness, acknowledgment of inequities, equal distributions of goods, and allocation. It is more holistic than how linear and binary policies around environmental justice tend to be. Indigenous communities do not advocate for just these things, because their cultural values are more complex than this linear way of thinking. Given that Indigenous communities have the highest *ecological debt,* Indigenous cultures continue to be threatened and impacted. I refer to ecological debt as experiencing the highest impacts of climate change and environmental degradation but not being responsible for causing these impacts or accelerating climate change. This is due to the reality that Indigenous peoples and communities are severely impacted by climate change and environmental degradation. This, coupled with the alarming murder rates Indigenous leaders of Latin America primarily face, means they focus their advocacy on everything from language revitalization to gender rights. This goes back to this Western desire to become experts of Indigenous cultures and communities but without understanding that Indigenous cultures are more complex than the linear and binary Western way of thinking that academia and policies continue to be deeply grounded in. With all of this information, we are left with the question, What is environmental justice for Indigenous communities of Latin America? Overall, environmental justice for Indigenous peoples in Latin America encompasses everything that is vital to them from language revitalization to water rights. It goes beyond just focusing on natural resources, but also what impacts and affects Indigenous peoples. Given that Indigenous peoples do not separate themselves from nature or their environment, they play a major part of the environment. Thus anything that impacts Indigenous peoples ultimately impacts their environments and everything that impacts their environments ultimately impacts Indigenous peoples.

Ecowars and the Displacement of Indigenous Communities from Central America

In 2014 and beyond there were several headlines that pointed to an "immigration crisis" that was taking place in the United States.[9] However, behind this immigration crisis narrative lies the stories that are often never told because they are the stories of Indigenous climate and postwar refugees. Due to nationalism being deeply embedded in each country, xenophobia transcends beyond the colonial borders that were established to create separate countries. Xenophobia is not present only in the United States as it is a transnational issue that overshadows the journey that many Indigenous climate and war refugees undertake in search of a better life. Within the transnational discourse of Indigeneities, it is important to ensure that I mention the xenophobia that is present in my communities and across the rest of Mexico. My father recounts his story as he immigrated from his country of El Salvador, made it to Mexico, and eventually the United States. His story demonstrates that sometimes the journey across the border is harder because of the constant xenophobia you are subjected to as an Indigenous person from another country. It is important to note that my father left El Salvador in the 1980s and it is infuriating and saddening to read recent ethnographic studies that point to the reality that the xenophobic narrative in Mexico against Central Americans has not changed but only worsened.[10]

> *Can you tell me a little bit more about the journey you took to make it all the way here to the United States?*
>
> I had suffered an injury from the bombardment of my guerilla camp and the on the ground battle that took place

[simultaneously]. So I was bleeding severely from my leg and could barely walk. I was using the tree piece I had found to use it as support as I pushed myself to walk. Since the guerrilla encampment had been dispersed from this area, the military was moving more inland. We were no longer in the Metapán region, but more upland, closer to the border of Guatemala. I do not know how long it took me to get to Guatemala. Upon arriving there, I was directed to a clinic by an elder who saw me walking. This clinic was providing medical relief to war refugees. There was another war taking place in Guatemala as well. Given that the country knew what war was like, because they were in one as well, people there were very supportive of my healing. I was able to get my bleeding to stop and my leg was bandaged up. When I healed and could walk a little more, I returned to El Salvador for my younger brother who was getting close to the age they were recruiting children to fight in war. So I knew I had to go back to get him.

So when you picked up my uncle, you both started your journey towards Guatemala again?

Yes, we both did. In Guatemala we interacted with a few people and then made our way to Mexico. While we made our journey through Guatemala, no one would stop us because they could tell we were not from there and also were not members of any armed forces. The way we were dressed and spoke Spanish gave it away, so it was an easier journey than the one we experienced in Mexico. When we made it to Mexico, we started experiencing mistreatment. It was until I made it to your mom's pueblo that a family saw us, two young boys, and offered us to work for them in exchange for food and housing. Even though it was hard labor in the fields, they were offering us food and somewhere to sleep, so we took it. However, not everyone in your mom's pueblo was welcoming of us. To them we were *fuereños*, and they constantly reminded us of that. They would ask us to leave, but the family that had supported us protected us as well. We continued to receive this mistreatment, and sometimes it was far worse as we made our complete journey through Mexico.

PHOTO 5.1: My father and mother in the Frida Kahlo exhibit. Photographer: Victor Hernandez Jr.

*How did the mistreatment you received in Oaxaca and through-
out Mexico make you feel?*

It made me feel like I was worthless. I mean I was young so
I did not comprehend where so much hate came from. I was
confused as to why people who did not know us hated us
and held on to this deep resentment against us. It also taught
me that despite sharing an identity, sometimes there will be
people who will not like us just because we are from a different country. It also made me resentful because I could not
catch a break from my life. I had just gone through war, but
I felt like I was fighting a different war in Mexico. When we
first made it to Mexico, the policemen saw us and asked us
where we were from. We said we were from Mexico, because
that is what people in Guatemala told us to say. However,
they decided to steal our shoes and the little money, *pesos,* an
elder had given us in Guatemala to get some food when we
made it to Mexico.

Nationalism has created a domino effect that recycles xenophobia.
This xenophobia alludes to mistreating others based on the sole premise that they are from a different country. Nationalism teaches people to
have pride in their country. However, this pride is only manifested within
the nationhood that was developed and created postcolonization. Thus it
exemplifies the significance of settler colonialism by honoring the settler
state and not the Indigenous lands and nations the Americas continue to
be. The United States as a nation has created xenophobic rhetoric against
people from Mexico, thus recycling racism. Mexico has done the same as a
nation and has created a xenophobic narrative against Central Americans.

It is important to mention that xenophobia is not only portrayed
through words but also actions. Sometimes this xenophobia, coupled with
anti-Indigeneity, makes Indigenous peoples from Central America victims

of structural violence they face as they make their journeys through Mexico in order to make it to the United States. Unfortunately, many do not make it through the United States because this structural violence results in experiencing extreme abuse, rape, dismemberment, and sometimes death.[11] It is sad to read, listen to other stories, and center my own lived experiences as someone who embodies dualities in her Indigeneity and identity, because I wish my communities could embrace one another. However, xenophobia separates and divides them both because of the structural violence that my Indigenous communities from Central America face as they make their journey through Mexico. In Mexico, Indigenous peoples sometimes become the *oppressors* of Indigenous peoples from Central America due to xenophobia, and it continues to sprinkle down through the rest of Latin America.

Xenophobia has unfortunately influenced Indigenous relations across the Americas because, the way settler colonialism is sometimes deeply embedded, it ends up impacting Indigenous peoples in different ways that may not always align with ancestral ways of doing things. I have found myself within interesting scenarios and instances where I am either forced to call out the xenophobia I have experienced or I have to remain silent and internalize it. I would be lying if I said that I had never faced xenophobic comments in Zapotec spaces, because unfortunately that is not my reality. I have often been told by Zapotec community members that their "families do not like Central Americans." While this statement may not be seen as harmful to them, it is harmful to me because this means that their families accept only part of me but not my whole self as someone who embodies an Indigeneity from Central America as well. I know I will constantly continue to face this and have to learn how to navigate it to ensure that I am not forgetting or dismissing part of myself and my communities that are oppressed due to the xenophobia that is present in the United States, Mexico, and the rest of Latin America. While my father was supported by some families in my mother's pueblo, not everyone was embracing of him or others who have made their journey through Oaxaca and throughout Mexico from Central America, and this also includes Indigenous peoples.

Oftentimes, as Indigenous peoples, we perpetuate oppression against each other due to how deeply embedded settler colonialism is in our society. Some of this oppression is internalized by the individuals who are facing it. In the case of my father's story, his journey through Mexico made him experience a legal form of violence as those who perpetuated all of the violent atrocities against him and many Central Americans are not breaking any form of law—immigration practices make it somewhat legal to mistreat those from other countries.[12] If the victim is Central American and the crime was committed by a Mexican person, the Central American would most likely face imprisonment and deportation because of their status as a foreigner and immigrant in Mexico. Immigration policies throughout the Americas are harsh and sometimes lead to physical mistreatment of others from different countries.

Unfortunately, Central America continued to face a lot of negative things, and in the years 2015, 2018, and beyond, more people from Central America decided they needed to leave their countries and form *caravans*. Caravans are when people decide to travel or take a journey together so that they can protect themselves better. Given all the violence Central American refugees face in Mexico, caravans have become a form of traveling Central Americans have organized around so they decrease the violence and harassment they face collectively. In 2014, there was a surge in the size of the first caravan that came from Central America, which included a large number of unaccompanied minors.[13] Many were sent by their parents and family members because they wanted to offer them a better life that they could not picture for them in Central America. Their journeys were not easy and there was a lot of separation from their families and parents in order to take this journey. Older siblings had to look out for their younger siblings as they made their journey through Mexico.

In 2015, I became an educator in Little Rock, Arkansas, because I wanted to serve students of color. This was the hardest job I have ever had in my young life, and it was not because it was demanding or a lot of hard work. It was the hardest job because I had to serve a lot of the children who came in the caravan of unaccompanied minors in 2014 and beyond. Many of them were able to be released from detention centers because they had an adult relative who could claim them and become their legal guardian. In Arkansas, the Central American population has grown and continues to grow each year.[14] Thus, it was no surprise that many unaccompanied minors made their final stop in Arkansas as they had family there.

However, within this unaccompanied minor narrative, the narratives of Indigenous children and adults continue to be dismissed. A lot of the children I served were not only Central American, but also Indigenous, mostly Maya. They were forced to leave their homes and families behind because of the environmental and violent crisis that overshadows Central America to this day. Once within educational systems many of them are placed in bilingual education, where they are not just learning one colonial language, English, but also Spanish. To me, this is how bilingual education fails to incorporate Indigenous children from Latin America because it centers two colonial languages, thus sometimes making it harder for students to learn.

I remember when one of my colleagues spoke to a student during lunch in Spanish. They were surprised that some students could not understand them when they spoke Spanish. It was known that people from Central America speak Spanish, so why weren't the students understanding them? This is why I have not only experienced how hard it is to learn multiple languages at once but also witnessed it as an educator. The students my colleague was speaking to were not just Guatemalan, but were Indigenous Guatemalan, which meant that their native language was not Spanish, but their native language. Now, in Arkansas, students are instructed only in English because of a law former President Bill Clinton passed when he was governor.[15] Thus, even if there are any bilingual initiatives that center Spanish, another colonial language, it cannot be implemented as thoroughly as in other states like California, where bilingual education is huge.

Language is one barrier that many Indigenous refugees continue to face as everything that is within the immigration jurisdiction, including services, is only offered in the language that all people from Latin America are assumed to speak, Spanish. This ignores how Indigenous peoples from Central America and other parts of Latin America oftentimes only speak their Indigenous language and not Spanish. Thus the services they provide in Spanish are not truly serving them.

After 2014, the number of people from Central America who took their journeys in caravans surged. This is due to not only the violence that war left in this region, which will take generations to heal from, but also due to water and land injustices they have faced, coupled with climate change impacts. It is known that *water is life* for Indigenous communities across the globe because water determines the life expectancy, abundance, and relationship between living and nonliving things. However, Indigenous peoples and rural people in El Salvador are facing issues with water contamination that jeopardizes their livelihoods. Water contamination is attributed to agricultural practices, poor sewage systems, lack of proper solid waste management, and mining and oil spills. Contaminated water reservoirs have led to the health issues many Indigenous peoples in El Salvador face. With more than half of the country living below the poverty line, El Salvador is experiencing alarming rates of water contamination deaths.

Other spills from extractive resource methods also contribute to the continued water contamination in El Salvador.[16] Between May and June 2016, nine hundred thousand gallons of steaming molasses contaminated two major rivers in El Salvador.[17] Many aquatic species were impacted, thus prohibiting Indigenous and rural peoples from fishing for sustenance. The slow response of the government and its enforcement to make the companies responsible is one of the major reasons why water contamination in El Salvador continues to be drastic.

Now in Guatemala, the water injustices Indigenous communities face are very similar. Guatemala is considered the second poorest country in Latin America due to the immense economic disparity that exists between the wealthy and poor. This economic disparity also leads to water access inequalities for Indigenous peoples, who are usually poor and reside in the rural parts of the country. With 51 percent of its population being Indigenous and most of them living in rural locations, the water supply and access to drinkable water in Guatemala are also limited for Indigenous peoples.[18] In Guatemala there are two common scenarios for Indigenous peoples:

1. There is no access to drinkable water, or

2. The access to drinkable water is often interrupted or not functional.[19]

Cities and towns in Guatemala do not receive the economic resources needed from the government to provide better water sanitation and services to the rural parts of Guatemala. While increasing the water services prices was an issue that was advocated for in the past, the opposition the community presented pushed this idea back. However, there has been little to nothing done to try to alleviate and fix the water issue that Indigenous peoples in Guatemala face.[20] Unfortunately, this impacts Indigenous children the most and is the reason why they face a rate of high malnutrition that ultimately impacts their overall health, well-being, and livelihood.[21]

The reciprocity between Indigenous peoples and water is essential for Indigenous communities in El Salvador and Guatemala, and due to habitat loss as a result of agricultural practices, spills, sanitation problems, and lack of services to rural locations, both Indigenous communities of El Salvador and Guatemala seek water justice—a form of environmental justice. However, due to the violent ways in which Indigenous leaders are treated and oftentimes murdered in Central America, communities are seeking refuge in the United States and Mexico. We cannot ignore the water scarcity, sanitation, and access Indigenous peoples from El Salvador and Guatemala face and not understand why this is also pushing them to leave their native lands.

Habitat loss due to agricultural practices also contributes to their water injustice. When we look at the refugee crisis that is occurring through an environmental justice lens, we can further understand how we can continue to advocate for the survival of Indigenous peoples. *Water is life,* and if they do not have access to it, their livelihoods are jeopardized and more barriers are created that prevent them from achieving environmental justice in the United States. The ongoing ecowars against Indigenous leaders and community advocates ultimately push many Indigenous community members to make sacrifices and leave their native and ancestral lands in search of justice, protection, and survival. As Veronica Montes, assistant professor of sociology and codirector of Latin American, Iberian, and Latina/o Studies at Bryn Mawr College, wrote, "Why are these people leaving their homes, risking their families' lives, and setting out for the US–Mexico border? The answer is complex, and each country in the Central American region has its own history."[22]

Given my own community advocacy that is motivated by my lived experiences and the embodied lived experiences of my parents, family, and communities, the reason why Central Americans, in particular Indigenous peoples, are leaving their countries is indeed complex, but it is due to a myriad of influences such as the civil wars that ended in 1992 in El Salvador and 1996 in Guatemala, the extreme environmental injustices they face that result in extreme poverty, and the lack of protection for their Indigenous and land rights that results in the deaths of Indigenous leaders and advocates.

Gender Violence's Role in the Ecowars

We discussed the role xenophobia plays within the ecowars that are severely affecting Indigenous peoples from Central America who decide to leave their ancestral homelands for the United States. Xenophobia is often left out of the discourses of transnational environmental justice and Indigeneity discourses due to the immigration policies and laws that govern all countries differently throughout the Americas. Another form of oppression and injustice I have noticed that is also left out or only briefly mentioned in the environmental and Indigeneity discourse is the gender violence that severely impacts Indigenous peoples from Latin America. Gender violence has severely impacted the Zapotec community of Oaxaca by putting Indigenous rights leaders and advocates who have fought to protect gender rights in danger.

In the Zapotec community we have a third gender that is esteemed and highly respected in our culture, and those who are of that gender are referred to as *muxes*. In order to understand how gender rights are interconnected to our environment, I interviewed an esteemed muxe leader in the Zapotec (*Binnizá*) community, Elvis Guerra.

What relationship do you have as a muxe with our environment and Mother Earth?

The relationship of the muxe with nature is extremely strong, because through traditions you learn to respect the natural elements. For example, in *la vela* (traditional party that takes place at night, dedicated to a deity, family, or saint whose commitment is with society and with God), *guelabe'ñe', vela al largarto* (reptile), is a ritual to venerate said reptile as an ancient God of the Zapotecs. [There is also the] *vela de la Santa Cruz,* Holy Cross of the Fishermen, where the sea is thanked for its enormous goodness, because for us, every time the fisherman throws his net, the Santa Cruz (Holy Cross) of the fishermen grants permission to use their fish, to enter the sea, an element so bleeding for us, or the Jasmine candles or *biadxi* (plum) to venerate the flora of the Binnizá, is very interesting because of the cultural syncretism it embodies as a result of

the Christianity we adopted the Santa Cruz as part of us and
nature.

Elvis shares how their relationship with nature as muxe plays a major
role in how they are taught to not just respect but also honor nature and
Mother Earth. In the Zapotec community, muxes are highly esteemed
because they embody a higher spirituality than that of cisgender men
and women according to our ancient codices. Muxes are mostly identified
during their childhood, and this is how the community has been able to
protect them and maintain this gender identity since time immemorial.
When a mother realizes that her son was born as a muxe, she makes the
arrangements so her son can get a muxe role model to teach him how to
maintain an integral role in the Zapotec society while being taught how to
navigate their identity. A muxe can choose to dress in regalia that women
in the community wear or decide to dress in the traditional clothing men
wear. However, a muxe that dresses in women's regalia is ostracized by the
patriarchal systems that have also been infiltrated in the Zapotec community, because patriarchy is amplified through homophobia. Relationships
that muxes sustain are condemned by Christianity and Catholicism, two
religions that are very present in the Zapotec communities, Oaxaca, and
the rest of Latin America. I continued my discussion with Elvis about the
role muxes play in the Zapotec society and how patriarchy impacts muxes.

What role do muxes play in the Zapotec society?

[Muxes play] an economic function because we contribute to
the economic validity, through commercial exchange . . . there
is a monetary remuneration, and because we continue to pay
for the family, for the care of the parents. A sexual function
because we muxes are the ones who initiate young people into
sexual life. A social function because we participate in the daily
life of Juchitan through our work as embroiderers, makeup
artists, hairdressers, cooks, etc.

How does patriarchy impact you personally?

Of course I was impacted by patriarchy. From my uncles who opposed my way of dressing, to my father who hated muxes. Later at school, when they wouldn't let me arrive dressed in skirts and a huipil, or refuse to be in the girls' line. As a result, I had to learn how to defend myself.

Elvis's story demonstrates the important role muxes play in the Zapotec community and overall Mexican society. Unfortunately, sometimes their role is not valued because of homophobia that overshadows their experiences and advocacy. Due to the gender violence that they face as well, many muxe leaders and advocates are murdered. This shows how even within the ecowars, those fighting for gender rights face violence and harm. One Indigenous leader who was a strong advocate for muxes was Óscar Cazorla López. He was a muxe activist for over forty years and was one of the founders of *Las Intrépidas Buscadoras de Peligro*,[23] a muxe group that hosts the annual muxe ball and gala where a queen reigns each year. There is an outdoor parade that takes over the entire city of Juchitan, and muxes dressed in their elegant regalia and jewelry walk in this parade. *Las Intrépidas Buscadoras de Peligro* is a sisterhood that continues to build a welcoming environment for the muxe community in Juchitan.[24]

Óscar did a lot of the behind-the-scenes work and preparation for this annual ball as he taught the muxes the dances and cultural protocols that are followed in this event. It has been active for forty-five years and it is important to mention and recognize that this ball has allowed our community to become more accepting of our muxes as religion was pushing away our cultural belief of being accepting of our muxes. Óscar was known as the patriarch of the muxe community, and it was thanks to him that muxes have more rights, a stronger voice, and are even accepted in other regions outside of Juchitan.[25] Óscar was assassinated at his own home, and his murder shocked the entire muxe society. Despite him being violently murdered, to this day the culprit has not been found and no additional information has been provided about his case.[26] Unfortunately his murder has not been the only one that has impacted the muxe community.

Óscar's belief was that providing education to the muxe community allowed them to become less vulnerable. This education that he led not only provided the muxe community with sex education to prevent HIV infections but also other educational opportunities that he could find funding for as a professional accountant. The ball he helped create was one of the ways Juchitan recovered financially because it attracted tourism and generated the economy that supported our region during the earthquake relief efforts.

He was found murdered on February 9, 2019, and continues to be missed by our community and his family.[27] He had a beautiful funeral, which demonstrates how loved he was in our community. He was given his final goodbye with *La Zandunga*, an important song in the Zapotec community. His strong activism continues to be a legacy that will continue to be honored in Juchitan by our muxe community.

Amitaí Verdugo, a muxe advocate and leader, explains how the muxe identity is changing to adapt to the present-day. Given the modernization that all of Mexico is also undergoing and the nuances that are being integrated within gender identity, the muxe identity is adapting as well. Amitaí Verdugo recounts her story.

What does the muxe identity mean in this century?

For me, the muxe identity in this century means more than a single word, a set of various actions, such as tolerance, love, respect and acceptance, today despite still living in a society in which patriarchy is very notorious, our identity as muxes is advancing and achieving a change, albeit slow, but we continue to fight for our duality and identity to cover more space in our society.

How has the patriarchy that Mexico continues to follow in Mexico impacted you?

Patriarchy in Mexico as a society is predominant and has really had a great impact within the muxe community, since it has even claimed lives, and in the new generations these actions have caused fear of opening up to society as muxes.

Will the Ecowars against Indigenous Peoples in Latin America Ever End?

This is a hard question to answer because I want to be hopeful that one day peace can be restored in Mexico and Central America so that Indigenous peoples from those lands are not forced to leave. However, that makes me feel like I am being unrealistic because settler governments that operate under capitalism only prioritize one thing, *capitalism*. Capitalism across the Americas was built on stolen Indigenous lands and stolen Indigenous peoples, mostly from Africa. As a non-Black Indigenous person, I am not positioned to speak on the Afro-Indigenous experiences of Mexico or Central America, but I believe Indigenous communities also have to constantly fight anti-Blackness internally and externally because settler colonialism also introduced anti-Blackness in the same way it introduced xenophobia and homophobia into our Indigenous communities. This is a part of the continued ecowars Indigenous peoples of Latin America face. In 2020, Honduras continued to face extreme violence enacted against their Garifuna Indigenous leaders and advocates. Mark Anderson, Professor and Chair of the Anthropology Department at University of California Santa Cruz, explains the Garifuna identity:

> Garifuna, according to scholarship produced by their own intellectuals and other scholars, are a people descended from marooned African slaves and Indigenous peoples of the Caribbean. They

were deported from the island of St. Vincent to the coast of Central America in 1797, eventually settling in present-day Honduras, Belize, Guatemala and Nicaragua.[28]

Unfortunately, the Garifuna community leaders and advocates continue to face extreme violence throughout Central America due to the anti-Blackness, coupled with the ecowars that continue to impact Indigenous rights and land defenders. In 2020, five Garifuna leaders in Honduras were kidnapped by armed police forces due to their advocacy in protecting their lands from the further exploitation drug traffickers and palm oil and tourism corporations continue to enact against them and their lands.[29] The following year, in 2021, two Indigenous Garifuna leaders, Martin Pandy and Víctor Martínez, were murdered.[30]

For the Garifuna people, part of their Indigenous and land rights revolves around dismantling anti-Blackness that is also deeply embedded in all of Latin America, including non-Black Indigenous communities. They, like other Indigenous leaders and community advocates, are not immune to the violence that is increasing in Latin America against Indigenous peoples. Settler colonialism introduced racial caste systems in Latin America that continue to marginalize and legalize the oppression of Indigenous and Afro-Indigenous peoples based on the premise that being Indigenous and Black means that they are at the bottom of these racial caste systems.

For me personally, in order to defeat the ecowars as Indigenous peoples, we must become a solidified nation, which means that intertribal and pueblo relationships must be fortified. However, we must understand that we are going against a giant system, settler colonialism, and we need folks who will be willing to give up the power and privilege they are granted within this system based solely on their race or socioeconomic status in order for us to truly reach environmental justice and end the ecowars. Indigenous peoples from Mexico, Central America, and the rest of Latin America face a lot of obstacles, and sometimes it is hard to constantly fight throughout our lives.

Sometimes I feel that it is my sole obligation to constantly advocate for my Indigenous communities, especially those who are still fighting these ecowars back in my native homelands. However, it is time that the hard work and sacrifices are not only done by Indigenous peoples, but also settlers and non-Indigenous peoples so that we can truly create just worlds where Indigenous peoples do not have to constantly fight these ecowars. I am ending this chapter with two questions to help you reflect and think of what you can do to help, because oftentimes the oppressor asks the oppressed for solutions to the oppression. This shifts the obligation to do the hard work toward the oppressed and not the oppressor.

1. What are you willing to do to help Indigenous communities across the Americas that continue to face the ecowars?

2. What role do you play in the xenophobia, gender rights violations, and anti-Blackness that are present through the Americas?

I hope that these questions help you as the reader identify some things you can do to help, support, and amplify Indigenous voices and rights across the Americas. I hope they also help Indigenous readers identify the work they have to do internally and externally so that we can fortify our relationships with each other.

6

Tierra Madre:
Indigenous Women and Ecofeminism

When all men learn how to respect all women,
they will also learn how to respect the most
powerful of them all, Mother Earth.

—MARÍA DE JESÚS, my grandmother

It is not surprising that for many of us, we can use the treatment toward our Mother Earth as a metaphor to how we are treated. Unfortunately, the femicides against Indigenous women continue to impact all of our Indigenous communities across the Americas. The same way human beings are destroying our Mother Earth, our spirits are being destroyed as Indigenous women. We continue to live under the notions of patriarchy that dictate that our social place should be below a man's political, social, and cultural power. However, time after time, we continue to bear witness to the movements Indigenous women are leading and the work they are tirelessly doing for their communities.

Unfortunately, because of patriarchy, Indigenous women's voices are silenced and ignored by mainstream media. For example, patriarchy, coupled with anti-Indigeneity, has formulated a stereotypical role for Indigenous women in Latin America. In Mexico, we see how these roles are subjected to how Indigenous women are portrayed in the media as well. One example that brings this to light is the character of the "India María." As Seraina Rohre, visiting scholar at the Chicano Studies Research Center of UCLA, describes this role,

> La India María—a humble and stubborn Indigenous Mexican woman—is one of the most popular characters of the Mexican stage, television, and film. Created and portrayed by María Elena Velasco, La India María has delighted audiences since the late 1960s with slapstick humor that slyly critiques discrimination and the powerful. At the same time, however, many critics have derided the iconic figure as a racist depiction of a negative stereotype and dismissed the India María films as exploitation cinema unworthy of serious attention.[1]

There are nuances to this character, specifically in regard to Mexican cinema because, while it might have been widely accepted in the 1960s throughout Mexico, the character stereotypes Indigenous women based on patriarchal notions. These notions dictate that Indigenous women should be and should remain uneducated, ignorant, and submissive and only take roles within the labor and service industry. Indigenous women are meant to be maids and not be intellectuals, politicians, or hold any power and privilege. These patriarchal and anti-Indigenous notions are rooted in settler colonialism, because in Mexico the racial caste systems continue to determine that Indigenous peoples are not within the systems of power and privilege. La India María is how patriarchy, coupled with anti-Indigeneity, continues to portray Indigenous women and the roles it grants them while failing to acknowledge how these stereotypical behaviors allude to the realities that settler colonialism places women under.

Educational opportunities continue to be limited for Indigenous peoples, in particular women in Mexico, because they are subjected to taking on household roles rather than educational roles. In my extended family, I am asked about my decision to not have children yet, as it is the norm for women in our society to have children in their twenties. I was the first in my ancestral lineage to pursue higher education as it was often not expected from Indigenous women or granted to Indigenous women in our societies. This is how patriarchy continues to be manifested under the threat that if a woman does not abide by what is the norm, she is subjected to punishment through violence and physical harm. The character of *la India María* also continues to define how "representational conventions in Mexican ethnographic cinema often depend on the consignment of Indigenous people to a national past that perpetuates the racist discourse of aboriginal intellectual inferiority."[2] Sometimes I forget that this is the reality placed on us as Indigenous women, because I am immersed in academic bubbles and taking on different battles within racist, sexist, and other systems that are in place in the United States academic realms.

However, when I go back home and am subjected to mistreatment by non-Indigenous peoples or Indigenous men, they become more uncomfortable upon finding out that I hold a doctoral degree in the sciences because the sciences are deemed hard due to how exclusive they are to those who come from better educational backgrounds and are not Indigenous or youth of color. Yet, one of the reasons why I pursued the sciences, in particular the environmental sciences, was because my grandmother always told me stories that amplified the role Indigenous women have with our environments. We have a closer connection to our environments because of our roles within the spiritual and sacred world. This means that no matter the systems of oppression we face, it is our inherited right to have a strong connection to Mother Earth.

PHOTO 6.1: From left to right, my grandfather, my grandmother, my uncle, and my great aunt. Family Archive. Photographer: Unknown

Patriarchal norms that have been set for women in general are also why Indigenous men continue to be given the credit for movements and work that Indigenous women have led. A clear example of how patriarchy manifests in the erasure of women from Indigenous-led movements is that of the Ejército Zapatista de Liberación Nacional (EZLN).[3] This was the social movement that allowed the Tzotzil, Tzeltal, Tojolab'al, and Ch'ol Maya pueblos, who renamed themselves the Zapatistas, to reclaim their autonomy and ancestral lands in Chiapas. On January 1, 1994, armed with weapons, they marched to the city of San Cristóbal de las Casas, Chiapas, and declared their self-determination to follow their own government structures and not Mexico's. This has been coined as the *Zapatista Uprising* and it brought to the forefront conversations about Indigenous sovereignty and rights in Mexico.[4] Unlike the rest of North America (United States and Canada), Mexico's Indigenous pueblos and communities never signed treaties. This means that there are no existing treaties within Mexico that Indigenous communities and pueblos can fight for. However, there are articles placed within the constitution of Mexico that grants Indigenous communities and pueblos some rights. Given that this is in the constitution, these same articles can be amended and removed with government approval. For the Tzotzil, Tzeltal, Tojolab'al, and Ch'ol Maya pueblos, they had experienced a lot of injustices under the premises that settler colonial governments continue to follow. A lot of their land was being privatized and sold to large agricultural companies that were further displacing them from their ancestral lands. As a result, they decided that they wanted to declare their self-determination to dictate their educational, social, economic, and political structures they wanted their pueblos to be governed under. Thus they took years to prepare for this takeover in 1994. This preparation required intensive military training, similar to the one the army forces undergo, but there was one evident exception in their preparation—Indigenous women were allowed to train in these forces because among the leadership, including their main commander and leader who was behind this preparation to declare their self-determination, were Indigenous women.

Comandante Ramona

Comandante Ramona was the leader of EZLN and her title as *commander* gave her the highest ranking of all.[5] She led the communities by helping organize them with secret meetings since the 1980s. Like her community, she wanted a better future free from the oppression the settler government of Mexico was enacting on her people. However, one thing that separated her from her community was something that she shared with other Indigenous women: she wanted to gain rights for Indigenous women as they were still forced to marry at a young age and were prevented from taking roles that would enhance their livelihoods outside of the role of being a wife. She brought to light conversations and ideologies Indigenous women had to constantly fight against:[6] first, the sexism we face within our own communities due to the patriarchy that has infiltrated our communities and has been adapted by our own people,[7] and second, the anti-Indigeneity racism we face under the premises that Mexico follows, such as that being Indigenous is something to be ashamed of, because we are treated like second-class citizens. Sexism coupled with anti-Indigeneity makes our experiences as Indigenous women unique in perspective as we have to not only fight for our rights outside of our community but also within our communities.

Comandante Ramona not only had the highest ranking but the most to do as an Indigenous woman heavily involved in the leadership of this movement. Looking at the EZLN allows us to further understand how patriarchy manifests in Indigenous women-led movements. It was not Comandante Ramona who became the face of this movement but rather Subcomandante Marcos. Now the question is whether the community deliberately chose this, because Mexico would not take an Indigenous women-led movement such as this one seriously or want to negotiate with them, or whether this was deliberately done by the media because they only wanted to pay attention to the men behind this movement.

Not only that, Subcomandante Marcos is not Indigenous but rather *mestizo*. Now *mestizaje* has been a complicated topic to dissect as it is oftentimes seen as a topic around *blood quantum* (ideology introduced to quantify the percentage of Indigenous blood people possess) due to how it has been previously taught in scholarship that has been presented to us from previous generations.[8] However, mestizaje is not an issue of blood quantum but rather a social, cultural, and political identity that acknowledges someone's Indigenous roots but also recognizes their Spanish heritage.[9] It is enforcing the dualities that come with the identities of both the oppressor and oppressed. This is the national identity Mexico has amplified and forcefully placed higher in the racial caste system it continues to follow, because it is, in a way, a settler colonialism ideology to eradicate Indigeneity by assimilation and interracial marriages. Unlike

Native American communities or First Nations in the United States and Canada, our Indigeneity is not defined by blood quantum or enrollment but rather lived experiences and strong kinships with our pueblos, communities, and people. Assimilation continues to happen in Mexico, and this removes people from their Indigeneity. In this case, Subcomandante Marcos was an educated mestizo who also attended the prestigious university of Mexico, National Autonomous University of Mexico (UNAM). His positionality as mestizo and a man allowed him to obtain that power through the representation he gained within this movement.

Comandante Ramona and the other Indigenous women leaders became his shadows, and this, as mentioned before, can be due to two important interpretations: one, that they knew a mestizo man was going to gather the attention of the Mexican government and be taken seriously, or two, that they were not given the opportunity to become the representation or face of this movement because they were Indigenous women. Despite this discourse, Comandante Ramona continued her advocacy for Indigenous women's rights that eventually led to the creation of the Women's Revolutionary Law.

In the Women's Revolutionary Law, Indigenous women had to advocate for rights that were granted to Indigenous men in their communities but were denied to them.[10] These ranged from having the right to receive a salary to sustain themselves and their families to having a choice over their childbearing, choosing how many children they wanted to have. The first law listed in this decree is interesting because it is the one that grants Indigenous women the right to join the revolution despite the other intersectionalities they embody (e.g., political affiliation, age, etc.). This is interesting because despite EZLN having Indigenous women in leadership roles, many women were prevented from joining. This demonstrates how patriarchy continues to be sustained and amplified, even among Indigenous women-led movements.

The Women's Revolutionary Law was published in *El Despertador Mexicano* (the Zapatista-led newspaper); therefore these granted rights and demands were circulated throughout the communities. The EZLN movement was not only about land liberation but also about Indigenous women's liberation. It is their resilience, pride, dignity, and passion that drive Indigenous women from Southern Mexico to achieve a better life and provide more opportunities for their communities. Southern Mexico—Oaxaca, Chiapas, Puebla, and Veracruz, collectively—has the highest percentage of Indigenous peoples. Seeing this uprising and movement for the environment and Indigenous women's liberation has inspired nearby communities and pueblos to also advocate for their rights. Indigenous land and women's liberation go hand in hand as one cannot be achieved without the other among Indigenous communities.

Comandante Ramona died in 2006 from health complications, but her legacy continues to inspire Indigenous women to fight for their rights.

Indigenous Women-Led Artisan Collectives

The Women's Revolutionary Law also advocated for Indigenous women's rights to receive salaries and become an integral part of the economy and workforce in their communities. This helped Indigenous women because they continue to be the artisans for many of their communities. In all Indigenous pueblos and communities, weaving and embroidering have been one of the ancient legacies that has been passed down through generations. This skills trade in our Oaxaca communities is often taught to our women and muxes. As a result, our weaving and embroidering has become an economic revenue that allows Indigenous women to play an important and vital role in their family's and community's economy. This is how our traditional embroidering and weaving have been maintained and sustained for generations. In our communities, the economic and educational opportunities that are offered to our people are limited, so we maintain this skills trade as it allows us to garner some wages. However, this does not prevent people from exploiting our artisans. We oftentimes are witness to plagiarism or exploitation that comes from high-end fashion brands.

PHOTO 6.2: Jessica closing her huipil. Photographer: Jessica Hernandez

Our huipiles carry our environments, and the elements embedded in them are collected from our local flora, fauna, and ecological histories. For example, in the Zapotec community, flowers native to our region are embroidered onto their huipiles. The flowers inspire us, and we as Indigenous women have a unique relationship with them as a matriarchal society; we are the flowers that bring strength and beauty to our communities. Not only do we incorporate flowers, we also incorporate geometric figures that resemble some of our ancient structures, such as our pyramids. This is how we maintain the past (our ancestors) with our present and future (descendants). My grandmother always told me that when we weaved or embroidered, we weaved our legacies and embroidered our resistance. Our huipiles, made with our embroideries and weaving styles, are our culture, and we can grant this gift to each generation. When you embroider a huipil, you take hours, you dedicate part of your life, and this is done with love and care. They are artistic interpretations of our environments, and this is why Indigenous women and muxes carry a deeper connection to our environments. It inspires our embroidery and, in other Indigenous communities, their weaving.

As my grandmother always told me, with a huipil we can carry part of our environment whenever we go. It allows us to create a replication of our environments because, like our environments, our huipiles also have a symmetrical feature. For those of us displaced, sometimes this is the only piece of our ancestral lands we can physically carry. Embroidering and weaving are unique skills Indigenous women have because these skills allow us to connect our nature to women. It is the women-nature nexus that is very present in our communities, and this is why it is important to protect our embroideries and weaving that go into our huipiles. By protecting them, we protect our Indigenous women, our self-autonomy, our liberation, and ultimately our environments that inspire them.

Indigenous women-led artisan collectives are another movement Indigenous women have led for the betterment of our communities. They liberate our communities from being exploited and from high-end fashion brands plagiarizing our work, and they allow us to continue fighting for Indigenous women's rights.

As Indigenous women, we continue to fight against patriarchy, but have to also fight against the plagiarism, appropriation, and exploitation we are forced to succumb to under capitalism that values our work as *cheap* but says it is *worth* more when a non-Indigenous name, like these fashion brands, are attached to them. Resellers purchase our textiles, embroideries, and weaving and sell them for an extreme amount that triples or sometimes quadruples their profit. We have to be honest and explain the haggling we experience as artisans, especially when we sell on the side of the road. I recall the many times I would accompany my grandmother to sell her embroidery and oftentimes she was haggled for her work. I could recall the sadness that filled her face when she was offered less than what she was asking for her products initially. There comes a time when we have to think about whether we will wait for a fair compensation or whether we will have to accept anything because we are sometimes in need.

PHOTO 6.3: Wearing my grandmother's huipil passed down to me during my master's graduation. Photographer: Victor Hernandez

Oftentimes, many of our Indigenous women artisans travel to the city of Juchitan, where they sell their huipiles and work for relatively cheap. There is a desire for cheap products that these collectives or fashion brands, oftentimes not Indigenous owned, want to make a huge profit from. Unfortunately, many other resellers continue to make money off our Indigenous women artisans. Oftentimes, we face exploitation by other non-Indigenous people in our countries who have the means to mass-market our products. It is important to acknowledge that our Indigenous pueblos and communities do not have access to high-speed internet, or cell phone reception, which is something our communities share with many Native American communities in the United States. We have resellers in Mexico who are selling our huipiles for a higher profit than what we are paid. This shows that the least appreciated skills trade is the weaving and embroidering we as Indigenous women do. This also calls for the importance of Indigenous women-led artisan collectives that allow themselves and other Indigenous women in their pueblos and communities to sell their embroideries and weaving at a reasonable price, without having to go through a third party that makes more profit than they do.

In order to understand the significance of huipiles and our environment, I received the following testimonial from Beatriz Montesinos Fernández, member and representative of the *Colectivo Telar Triki*.

What diffusion do Triqui textiles have in relation to the environment?

Our textiles arose from our need to dress to cover ourselves from the weather. We take elements of nature to be able to make our clothing. Our grandmothers used cotton and wool and transformed it to be able to create with their hands the textiles that would serve the needs they had at that time. We take advantage of the knowledge that the grandmothers left us, and we continue to make our textiles to cover ourselves. But I think that like all human beings, they had the need to transmit their knowledge and they were able to translate it into their textiles; that is why they also taught us to express ourselves on our loom. Thanks to our grandmothers we now know that our loom is a blank canvas in which to capture all knowledge— knowledge of our grandmothers and also of ourselves.

How is the environment reflected in textiles?

Our textiles are the reflection of our environment, because we continue to shape the animals, plants, and objects of our environment in our huipil. As children we have been taught to admire the loom, and indirectly we have been taught to express ourselves in it. As we sit next to a woman who is weaving, she will always talk to us about the meanings and mysteries that a huipil keeps. In this way, learning to weave a loom begins to draw attention to girls. This is how we learn to express ourselves at the loom. Although I went to school, I did not feel the freedom to write; I liked more to mix colors on the loom. I think it is also a way of expressing yourself. Our huipil is made of many butterflies, our grandmothers chose an animal from nature to express themselves, because in our huipil you can read dozens of different butterflies. Butter- flies are sacred animals, and butterflies are a part of the fauna that is important to us; butterflies are the natural phenomena around us. I think that the huipil that I wear is already full of natural writing.

What is the mission of the project that the collective you are a part of upholds?

This collective was born mainly from the need to pay for our financial expenses. We began to make ourselves known through digital platforms because for us it is very difficult to market in a store in the city or go to offer our products in the cities. We have looked for people who have supported us and we want to continue advancing in this collective so that our work gives us economic support. We also want to continue growing so that we can invite more women. We want to offer our textiles to the public without intermediaries, since we have seen many cases in which people from the cities come to win with our work. We want to be the ones who dictate our business and our profits.

The Triqui community is located in Northeast Oaxaca and is known for its beautiful long red huipiles. They hold strong agricultural knowledge, which is one of the reasons why they make up a large percentage of agricultural workers in the diaspora, especially in the United States.[11] Like most Indigenous communities of Latin America, their community has unfortunately faced extreme violence, in particular their women leaders. In 2009 their pueblo in San Juan Copala was blockaded by the Mexican military. Over twenty-nine people, mostly women, were killed as a result of this siege, including community advocates and leaders, Beatriz Cariño and Jyri Jaakkola.[12] Jyri Jaakkola was a human rights advocate and observer from Finland who was helping the Triqui community and women advocate for the human rights violations they faced.[13]

Despite everything Indigenous women continue to face in their communities due to settler colonial violence, they work hard toward their existence, resistance, and resilience. My grandmother always shared with me that through our huipiles, we are also able to share the stories that negatively impact us because that is the reality of Indigenous peoples. We are able to find the positive in everything and despite facing violent atrocities committed by the Mexican military or government against us, our huipiles are a way we can also escape our harsh realities and seek nature and our environment as a refuge. She told me that sometimes this was why we focused on expressing our feelings and entire selves in our huipiles by integrating part of our environments, because this served as our own sanctuary. It reminds me of my father's story as he also sought nature and his environment as a sanctuary during war. As I recount later, he mentioned that nature served "a protective role" for him during the war. It also served as his refuge when he wanted to escape his harsh realities while fighting in the war as an eleven-year-old child in his country of El Salvador. This is why nature is also important to Indigenous peoples, because it allows them to maintain their hope and spirituality and fuels their resilience, even during hard times. This is the power nature carries and continues to hold for us Indigenous peoples across the Americas.

Beatriz shared how her grandmothers were able to express themselves in their textiles. Beatriz was missing this in her educational career as writing in Western academia is more rigid, dictates a certain structure, and is limited on the expressiveness one can share. I have found in my own experiences that books are the only method of writing that can allow us as scientists and academics to express our thoughts and stories. Unlike peer-reviewed articles, which mandate that we write in third person and follow a linear structure, books can be tailored more to Indigenous ways of knowing and teaching. Like our huipiles, writing books about our own stories is a way we can express ourselves. However, huipiles are able to integrate visual elements of our natural world and are better for showing how expressive Indigenous women are. Nature and our environments inspire our Indigenous artisans to re-create their surroundings in their artistic way of embroidering, weaving, and other artistic ways.

I often show my students in my environmental classes photographs of different Indigenous huipiles from Latin America and ask them to tell me how the huipiles are interconnected to nature. I follow this activity with a question on whether humans are therefore separated from nature as most Western environmental sciences have taught us. This allows them to visually analyze that Indigenous women are deeply connected to their environments and that despite sometimes not being offered the opportunities to learn how to read or write colonial languages that govern their current countries, they can express themselves visually. For me, being able to represent ourselves visually is similar to the ancient writing our ancestors practiced through our codices. Their writing system was made up of figures and symbols, not letters like we have now integrated into many languages. Through our huipiles you can see the vivid colors, animals and plants, materials that are appropriate for our climates, nature mimicry, patterns that resemble our environments, and many other aspects that demonstrate the kinships we as Indigenous peoples have with our environment and landscapes.

Manos del Mar Oaxaca

Another Indigenous artisan collective led by Indigenous women is Manos del Mar Oaxaca. Manos del Mar Oaxaca was founded by an elementary school teacher, Ana Laura Palacios Cepeda, who wanted to provide opportunities to the Indigenous women artisans in her Ikoots community, a coastal Indigenous community of Oaxaca. The Ikoots community is located in San Mateo del Mar, and this collective was established after the 2017 earthquakes that devastated Oaxaca, Mexico, primarily Indigenous communities. In 2016, Mexico experienced several earthquakes, one in Mexico City and one in the Oaxaca and nearby regions. However, during this time all the governmental and international aid that was sent for earthquake victims was redirected to Mexico City, while Southern Mexico was completely ignored. The history of the earthquake that devastated Mexico City in 1985 was revisited, while the colonial history that continues to impact our Indigenous pueblos and communities was never mentioned during this time. The 1985 Mexico City earthquake devastated the entire country because a natural disaster preparedness plan was not created yet to manage earthquakes and the buildings throughout the city were not built to withstand earthquakes.[14] On top of this, natural disasters severely impact the victims' mental health, and given all the lack of resources and aid, many were traumatized.[15] It is important to note that many of these victims relived that trauma when they experienced the 2017 earthquakes.

However, with the 2017 earthquakes, Mexico City was not as severely impacted as Oaxaca, yet most aid went to Mexico City and not to our pueblos in Oaxaca, where most Indigenous communities reside. My aunt and relatives who lived through the impacts of this earthquake recount how all they received was a plastic bag, weeks after the earthquake took place, that was filled with junk food and one plastic water bottle. Unfortunately, to this day, many families in Oaxaca and nearby regions are still recovering from the 2017 earthquakes. Unfortunately, in 2020, Oaxaca experienced another earthquake on June 23. This earthquake impacted both the communities that were still recovering from the 2017 earthquakes and other new communities as it was a 7.4 magnitude earthquake. The impact and destruction left many of our communities houseless as their houses were completely destroyed or part of their homes, like their roofs, fell.

Ana, founder of Manos del Mar, witnessed how the disparities were amplified in her pueblo during the 2017 earthquakes, so she utilized her free time after teaching every weekday to establish this collective. This passion was driven by the discussion that was taking place in her community to find a sustainable model to continue selling their textiles without depending on the government, as the government failed them and continued to fail them when their community needed aid. Thus she wanted to create a collective that not only focused on the economic aspects but also the social aspects. In Mexico, Indigeneity is ridiculed and our Indigenous children are taught to be ashamed of being Indigenous because the caste system racialized Indigenous and Black people in Mexico as inferior to the other races, such as *mestizos*, that Mexico embodies as a now-multicultural country. Growing up to be made to feel ashamed of her Ikoots identity, she wanted to find a way to support the Indigenous children of her community.

Unlike other artisan collectives, Manos del Mar is a socioeconomic project that not only supports Indigenous women artisans but also builds a reciprocal relationship within the community. It also supports elders during these times especially by providing them with *canastas Ikoots* (Ikoots baskets) that are filled with traditional foods like fish, shrimp, and cleaning supplies that are essential during the COVID-19 pandemic that began in 2020. Ana has been able to support these social and cultural projects through mutual aids, which I will discuss further later in this chapter. I interviewed Ana, founder of Manos del Mar Oaxaca, and she shared many important insights with me.

PHOTO 6.4: Ana Laura Palacios Cepeda. Photographer: David Palacios

As an Indigenous woman, what role does your gender play in the work you do for your community?

Throughout time the management of economic activities has always been seen as a man's role. However, things have changed, but this does not mean that it has been easier for me. I think that coordinating a project has empowered women like me to pursue their goals and pursue their education. It has made us think outside of what we have been accustomed to, such as getting married at a certain age, having a family at a certain age, when in fact there is a great panorama to these antiquated customs. The artisans today know and recognize that as women, we are capable of leading a family, not just having one. I think that the Manos del Mar project has been a space to understand that dreams can be achieved no matter our gender and position in society as Indigenous women.

How has patriarchy impacted the community work you have done with Manos del Mar?

We have suffered first for being a woman, then for being relatively young. At the beginning of our work, they used to deny our food baskets (that we started providing during the pandemic) and our workshops for the children, but that's when you have to persevere and show others that you can. And it is not just a political action, but a community-based social action that seeks to transform positively our communities that are provided with very few opportunities to shape and empower our identity as Ikoots. Today I can tell you that Ana Laura, the coordinator of the social activities of the project, has been required to remain strong even when confronting the criticism of people (mostly men) about my actions as an Indigenous woman. But when you have an identity, a purpose, when you are part of the community and pueblo, that gives you strength to do things, because it hurts to see what your community has become, and I think these small actions are like grains of sand that help our community create a better society. A person who is far removed from our people and pueblo has a difficult time understanding our culture, our social perspective in which we as Ikoots people and a strong pueblo are immersed in.

What relationship do you have with your environment, in this case the ocean and coast?

My relationship with the ocean is sacred. In it one finds the strength, the ocean is synonymous with how you feel. When you feel bad or sad you must go to the ocean and soak in it to leave everything bad and throw it into it. The ocean for the Ikoots people also represents our goddess (siren) who is the one who blesses and takes care of us. So for me the ocean is synonymous with respect. And although men are those who are immersed in the sea due to the economic activity of fishing, it is the women whose hands transform any catch into food and are the ones who play a major role in commercialization of the seafood men in our families catch.

Ana shares her struggles as an Indigenous woman but shares how her passion and love for her community are what fueled her persistence. Oftentimes, we fail to recognize how ageism also impacts young Indigenous women in and outside of our communities. I can relate to this because, as a young Indigenous woman, I am treated differently in many spaces when my academic credentials are not brought to the table. People in environmental spaces either treat me as ignorant or lacking experience to be in these spaces. I recall one time where I was interviewing for an advisory environmental board that is appointed by the government as it works within the state level. The current board members have to select you as the candidate for consideration and push your candidacy to the government's office. I was not informed that they were looking for someone with decades of experience in the environmental field until after I had invested several hours meeting with all the board members individually. This board was interested in reaching out more to communities of color, my communities, yet they felt I did not have enough decades of professional experience, despite being a person directly involved and a member of those communities they were aiming to serve.

Oftentimes, we as young Indigenous women are brought to the table to have conversations, but these conversations end up being more extractive rather than leading to forming reciprocal relationships. Our young age is characterized by not having enough experience, despite the need for lived experiences to be incorporated and oftentimes weighted more than academic or professional experience.

PHOTO 6.5: Ana's grandmother weaving. Photographer: Ana Laura Palacios Cepeda

This experience taught me that during these times, environmental agencies have adapted politically correct language to ensure they do not face any scrutiny for being exclusionary of people of color, but in reality, their actions continue to be exclusionary. Being a young Indigenous woman adds layers that both Ana and I have experienced that continue to require us to work three times harder in order to be deemed responsible enough, experienced enough, and worthy of being in such positions.

Ñaa Ñanga Tijaltepec

Ñaa Ñanga Tijaltepec is another Indigenous women-led artisan collective that has a special history. It is a product of community-based work that a nonprofit led in the pueblo of San Pablo Tijaltepec, Oaxaca. For over four years, CADA Foundation, a nonprofit organization located in Oaxaca, Mexico, collaborated with Indigenous women from San Pablo Tijaltepec. CADA's mission is to foster and facilitate local artisans and suppliers in becoming part of the local and global economy. Through a community-based social design, the foundation provided artisans with tools so they could revive and maintain the traditional techniques present in their clothing.

Lita Ser, founder of the collective, explains how the CADA Foundation taught them about creating a small business model so the collective could be in control of who they sold their embroideries to and for how much. There are not many business models where you can purchase textiles directly from the artisans as they also lack technological tools such as cellphones and internet access. Through the reinforcement of their cultural heritage, they were provided with tools to become self-sufficient in embroidering their designs while creating a social market where they could not only rely on resellers but also sell their embroideries directly to the consumer. Today, thanks to the social exchange that took place between CADA Foundation and the group of Indigenous women, the Ñaa Ñanga Tijaltepec collective was formed and continues to thrive. They are an independent, organized, and trained collective that fosters collaborative work, structures, and business that allow the Indigenous women artisans to maintain their livelihoods and support their communities.

Ñaa Ñanga is the Mixtec word that translates to "mujeres de los juguetes" or "women of toys." They gave this name to their collective because, as Indigenous girls, the toys they were provided were needles, yarn, and the other materials required to embroider. While embroidering is a skills trade for the Indigenous women in their communities, they all have beautiful memories of their childhoods as they were enjoying and having fun learning how to embroider from their grandmothers and mothers. The collective's name is a beautiful metaphor of how we as Indigenous women are raised in our communities and pueblos. Our other toys are from our Mother Earth, as we are taught to play with our natural elements as opposed to factory-manufactured toys. This is due to the poverty we grow up in and also because this is an important component of our identity. We are not raised playing with toys but rather use rocks, twigs, and other elements we can find in nature to play with in combination with our imagination. My mother has always shared stories of her childhood and how she used to play with dragonflies and grasshoppers. Her sisters and she would try to follow dragonflies, and this was how they played with nature. Grasshoppers are also Oaxacan traditional foods, so catching them was also made into a game for children. However, in recent times, dragonflies have become a vulnerable species in Oaxaca due to climate change and other environmental impacts that have decreased their populations.

These stories allow us to bring a different perspective on how capitalism has now impacted our Indigenous communities. Yes, toys have become a multibillion-dollar market, but for Indigenous children, they are not as entertaining as natural elements or supplies that they can use to learn how to embroider or weave. Most of our children in our communities and pueblos are also happier with receiving school supplies over toy cars and dolls. This demonstrates how Mother Earth does not only become our first teacher but also our playground.

For the women of Ñaa Ñanga Tijaltepec, they continue to embrace their childhood memories in the work they currently do. Their work also allows them to have autonomy over who they sell their embroideries to. The Mixtec or Ñuu Savi community has faced a lot of exploitation by fashion brands that use their embroideries and make them into high-end fashion. These high-end fashion brands are also Mexican brands that target tourists and the high class of Mexico as their dresses or garments are sold at high prices. Thus the San Pablo Tijaltepec community has held town-hall meetings to discuss having all artisans sell their embroideries at the same price to prevent other artisans from being exploited or haggled by resellers or fashion designers that incorporate their embroideries into their clothing. This has helped them somewhat, but there is always the need to provide food for their families, so sometimes they do have to sell at a lower price than their entire community agreed to. However, this is discouraged by their council as it ends up creating a system of exploitation of their Indigenous women artisans.

Their embroideries tell the stories of the cultural resistance Indigenous women have led. As mentioned previously, many of the previous generations utilized codices to revitalize their ancient writing styles and embody them in their embroideries. Like most artisan collectives led by Indigenous women, Ñaa Ñanga Tijaltepec not only focuses on their small business model but also on community-led work to support families, especially single mothers and elders. I asked Lita to tell her story of what inspires her to do this type of work that centers Indigenous women and artisans.

What inspires you to lead the Ñaa Ñanga Tijaltepec collective?

What inspires me to do this work is the strength of the Indigenous women who are a part of this collective. Witnessing and listening to their stories on how they embody strength despite everything life has thrown at them, including all the problems and obstacles they face. A lot of the Indigenous women in our collective are single mothers or widows. A lot of them have husbands who had to migrate and leave their families behind. This has left all the responsibility to care for their children on them and despite it all, they still find the strength to contribute to our pueblo's economy and self-autonomy. From feeding the animals, taking care of the milpas, among other things, they find the strength to do it all. I see through their strength my commitments and responsibilities as an Indigenous woman who had support from her parents to leave my pueblo and obtain an education. Being able to leave my pueblo and come back has allowed me to understand that I can support them with the minimal tools I received through my education. Their strength is contagious, and this is what continues to fuel my inspiration to continue fighting for the new generations. We all wish and desire that all of the children in our pueblo are given better opportunities in life than we were; we want to be able to collectively send them to college to also decrease the need for our people to have to migrate and leave behind our ancestral homelands.

Lita has always mentioned that Indigenous women in her pueblo hold a strong relationship with nature and their environments because they take care of the harvest, plants, and also take care of the animals. She finds motivation to do the work she needs to lead her collective in the strength the Indigenous women exude through their responsibilities and commitments to not just their families but entire communities. They are inspired by the roles they have in their communities to also transmit those relationships into their embroideries. In 2020, since the heavy rains from the hurricanes destroyed their milpas, they expressed their ecological grief from this by embroidering their *nuni* (corn). This shows the immense relationship Indigenous women have with their environments and why it is important to continue protecting and advocating for Indigenous women's rights.

Ecofeminism: Indigenous Women's Rights

Given the strong relationship Indigenous women have with their environments, it is important that Indigenous women are brought to the forefront on environmentalism. As societies throughout the Americas, we must advocate for the recognition of Indigenous women as movements are not easy to lead and require a lot of work, time, and dedication. This is why it is important to start lifting and supporting Indigenous women more, something that Indigenous men also have to do. Ecofeminism looks at the women-nature nexus and, through the testimonies and analysis provided in this chapter, we can bear witness to the ecofeminism embedded in Indigenous communities across the Americas. Ecofeminism aims to interpret relationships women have with their environments and explore how these can be brought to the forefront of environmentalism. In a way it is the antithesis of the patriarchy that exists in environmentalism. This patriarchy is why men are the only ones uplifted as the knowledge holders, seekers, and scientists when it pertains to our environments. In many Indigenous communities, like the Zapotec, Mixtec, and Triqui community, it is Indigenous women who hold more knowledge about their environment and nature as women are the nurturers of our people, environments, and communities. This nurturing is not only focused on childbearing but also supporting youth, supporting other women, and supporting our matriarchs—elders. Women in our community have a reciprocal relationship with nature. This relationship does not stay only within us and our environments but it is also expressed through our traditional huipiles, textiles, and embroideries. As a result, Indigenous women deeply feel any loss to our environment as it ultimately impacts our future generations. These deep feelings are why we continue to be the leaders in many environmental, climate, and food justice movements. This is also what fuels our responsibilities, commitments, and artisanal skills that are not only related to our communities and families but also our environments.

Ecofeminism asks for respect toward our ancestral knowledge and way of life, especially the knowledge Indigenous women carry. As Indigenous women, we do not seek to be treated better but to be granted the same rights and treatment men are granted without having to advocate or fight for those rights. For me, ecofeminism is an integral component of my Indigeneity. This is why as Zapotec women, we also braid our hair in the style of a crown. This crown is supposed to connect us from the earth to the skies. It is also meant to resemble the sun that nourishes us, our crops, and is one of our gods. It also resembles the flowers that we are within our communities and matriarchal society. My grandmother always told me that this was the crown our environments provided for Indigenous women, and this is why our braids, and the *trenza de reyna,* queen's braid, is how we braid our hair.

It is no surprise that it is our Indigenous women who get to wear traditional clothing and regalia that are more detailed and adorned while the men's traditional clothing and regalia tend to be white and simple. As my grandmother would tell me, this is due to the optics that are meant to make our Indigenous women and muxes shine more. According to physics, white projects light and this is what our men are supposed to do. They are supposed to project the light we Indigenous women and muxes radiate due to our strength and close connection to our environments. This light is important to project to the skies because that is where our gods, goddesses, and ancestors reside, in the clouds. In my Zapotec community and culture, we believe that our ancestors reside in the clouds alongside the gods and goddesses that created us. This is why we are considered the "cloud people."

PHOTO 6.6: My nephew, Jonathan, and classmate, Valeria. Photographer: Dalia Villatoro

While our traditional knowledge systems are often invalidated or considered pseudoscience by Western science, the fact that men wear white while we wear colorful clothing is the unknown physics that is also embedded in our communities. As Indigenous women, we have to face both anti-Indigeneity and sexism, and through ecofeminism we can address some of this racism and sexism. Our leadership and heart work behind the services we provide to our communities need to be highlighted and brought to the center of the environmental, climate, and food justice movements.

7

Ancestral Foods: Cooking with Fresh Banana Leaves

ooking with banana leaves has been an integral part of both my Oaxacan and Salvadoran Indigeneities. Tamales from both regions have been one of the ways my families and communities have incorporated banana leaves into their traditional and ancestral foods. Tamales made in *hojas de platano* (banana leaves) symbolize our resistance as Indigenous peoples, because no matter how our landscapes have been altered due to ecocolonialism and climate change, we find a way to continue nourishing ourselves, from mind to spirit. For example, banana trees are not native to our regions. They are mostly native to Southeast Asia and northern Australia and were introduced into the Americas, primarily Mexico and Central and South America, by settlers (Spaniards and Portuguese) during colonial times.[1] However, we have been able to learn how to incorporate bananas and its leaves into our traditional foods. Being able to adjust our diets to environmental and climate changes demonstrates our quick ability to adapt while also protecting ourselves, our communities, and our people. This also highlights our abilities to understand our environments and enhance our knowledge that also includes plants that are not native to our region.

Bananas, like other nonnative plants (e.g., mangoes, pineapples, etc.), play a major role in Latin America in general. For example, Latin American countries contribute to the high imports of bananas among other crops globally. This is due to the many fair-trade agreements Latin American countries have signed with several countries. For example, Mexico signed the North American Free Trade Agreement (NAFTA) in 1994, which reduced trading barriers and eliminated tax tariffs on imports and exports.[2] This inspired the United States to attempt to create trade relationships with the rest of Latin America, but Venezuela, Argentina, Bolivia, and Brazil opposed these negotiations in 2004.[3] Consequently the United States agreed to move forward with the Central America-Dominican Republic Free Trade Agreement (CAFTA-DR) instead.[4] Due to all the fair-trade agreements, in 2019 eight Latin American countries were among the top fifteen exporters of bananas globally.[5] The privatization of nonnative crops within agricultural infrastructures has led to Indigenous peoples being displaced from their ancestral lands as well. In Oaxaca, due to the region's topography, how it ranges in its physical and environmental features, only 9 percent of the land is arable.[6] Arable land is land that can be cleared for agricultural practices like farming. However, most arable land is purchased by large agricultural corporations since they pay less wages to laborers in other countries and can purchase land cheaper. In Latin America, it is also estimated that the number of exports will be rising throughout the years. Unfortunately, these fair trade agreements that Latin America has signed throughout history end up impacting small-scale farmers and producers.

My family in Oaxaca has sustained a small-scale mango orchard and has seen the decrease in revenue from what they had generated in the past. Mangoes are also not native to our regions, but rather Southeast Asia, primarily India.[7] Like bananas, they were also introduced during colonization and have taken over major acres of land throughout Latin America due to the large agricultural corporations that have purchased hundreds to thousands of acres of land. My aunt Virginia who has been the main caretaker of our family's small orchard explains how this has impacted her and my small Indigenous pueblo in Oaxaca.

> ***Now that crops and many fruits are exported from Mexico to the rest of the world, how has this impacted the family's mango orchard?***
>
> For us personally, our great-great grandparents were able to use some of our ancestral lands to create this small orchard of mangoes, when they were introduced to Oaxaca. They were working outside of the pueblo in larger mango plantations and then came back to our pueblo to teach other families so that they could create a new source of income, as economic opportunities for Indigenous peoples is limited in Oaxaca. However, it has been hard to continue competing with large-scale mango plantations because they are the ones who are able to meet the demand by having the ability to supply tons of mangoes that many of our pueblo's family-owned and communal mango orchards cannot compete against. As a result, we focus on selling our mangoes in our *tianguis* (open-space outdoor market). However, when we really need to make some money, we end up selling our mangoes to the larger corporations who operate larger plantations in our region for a really low price. This requires us to travel with our harvest to these plantations. They really take advantage of us since they know they are often our last resort.

While my family's small mango orchard has survived, many of our other pueblos' family-owned orchards have not because these fair-trade agreements that have been enacted in Mexico and Central America, as well as other parts of Latin America, have opened the country's market to large agricultural corporations that are usually owned by foreign entities that can buy large masses of land to create their plantations. Our communal orchards, which include bananas and mangoes and other nonnative crops, like pineapples, are also suffering due to climate change. Sometimes there are dry periods from the droughts or heavy rains from the hurricane season that end up harming not only our native agricultural systems, like our milpas, but also our adaptive agricultural practices, like our small orchards. While larger agricultural corporations have the means to provide water to their entire plantation or build canopies or use other equipment to cover them and protect them from heavy rains, unfortunately, most Indigenous and small-scale farmers do not have the same means. My pueblo has lost several of their communal harvests from their milpas and small orchards due to extreme droughts or heavy rains. On top of that, climate change is also introducing diseases (e.g., Panama disease that impacts banana trees) to many of the crops we communally take care of.

In Oaxaca we cultivate 2 to 3 percent of all the bananas that are harvested in Mexico, and we normally cultivate six different *plátanos*.[8] These include *plátanos pera, morados, manzano, machos, dominico,* and *cavendish*. Unfortunately, due to climate change, Panama disease, which derives from a soil-inhabiting fungi, continues to thrive in these conditions, impacting many plantain fields located in Oaxaca.[9] This fungus, which primarily impacts *plátanos machos* and *manzano* in our Oaxacan regions, blocks the banana trees' vascular systems and prevents them from receiving the proper water and nutrients throughout its system, causing the trees to wilt.[10] While some plantain fields can recover, infected banana trees have to be quarantined, sanitized, and closely monitored. Climate change is one of the culprits causing such plant diseases to manifest in regions where they were never found before. Drastic changes to any climatic conditions help fungi that cause the disease to thrive, thus resulting in drastic impacts to our flora, including both native and nonnative species. However, like us, the banana trees are resilient and have undergone various changes that include both their overall health and their displacement.

All of these bananas or *plátanos* (as we refer to them more commonly in Spanish) are different in taste, texture, and how they should be consumed.

Some of them are sweeter, like the *plátanos morados,* so you can consume them like most fruits while others oftentimes need to be cooked or fried, like *plátanos machos.* In El Salvador, we refer to them as *guineos*—green plantains that have yet not ripened. Bananas have always been one of my favorite fruits growing up, and my father would always tell stories of how our people are like bananas. They may be from the same plant family in plant classifications, but they are all different and unique in their own way. Even within the same species of bananas, not all of them are identical to one another in flavor, texture, and shape. Among Indigenous communities, we may be from the same pueblo, tribe, or nation, but we are all different and unique in many ways. This continues to solidify the importance of ensuring that we do not create or apply monolithic discourses onto Indigenous communities. Of course, humans are more complex than our plant relatives, but we all embody intersectionalities that make us different. Not all bananas in a bunch will be identical, and this applies to Indigenous communities as well. For us in the diaspora, we, too, are displaced from our ancestral homelands like banana trees were displaced, but we adapt to the new environments we find ourselves in. Eventually, we all become an integral component of our new environments like banana leaves and *plátanos* have become in our traditional diets across Latin America. Cooking with fresh banana leaves and making tamales from both of our regions, Oaxaca and El Salvador, allow me to see my history and ancestral ways through the banana leaves. They also allow me to reflect on the dark history behind the large corporations that export bananas and other fruits (e.g., mangoes, pineapples, etc.) from Latin America to the rest of the world.

Banana Republics

My family was displaced because of a war that was not just a civil war but a genocide that took place against our Indigenous peoples of Central America.[11] In 2013, Guatemala was able to convict a leader, Ríos Montt, who was the political actor that led the genocide committed against the

Indigenous peoples who lived there.[12] I think that outside this civil war narrative that occurred in Central America, we often forget to understand the implications of it just not being a war, but a genocide committed against Indigenous communities of those countries. I personally have witnessed the huge decrease of my Maya Ch'orti' community in Guatemala, Honduras, and El Salvador as our populations plummeted during the war. We are a border community because, due to border separations and creations, our nation was separated among three different countries: Guatemala, Honduras, and El Salvador. Our communities understand the position we have been placed in, and I think this is one of the major reasons why my father was helped in Guatemala when he left El Salvador.

Presently, one of the main focuses in our Maya Ch'orti' community is to revitalize our Indigenous language because, through genocide and war, our Indigenous languages, cultures, and traditions have been jeopardized and fractured.[13] Our Maya Ch'orti' community experienced a huge loss of our fluent speakers of the language as they were either killed during war or they are holding on to the trauma they endured through their language. I have witnessed in my life how our Indigenous languages oftentimes carry the trauma our people endured, just as our environment and lands do. As a result of the trauma, coupled with the stigma of being seen as less than for being Indigenous and speaking an Indigenous language, many generations choose to not pass down their languages to the younger generations. For my family, through the language, the genocide they endured is still very present, and it is hard for the older generations to speak their language because they have not healed yet.

Oftentimes, when we talk about the genocide against Indigenous communities, we contextualize it as something from the past, something that occurred hundreds of years ago. However, for Indigenous peoples of Central America, including my family, this genocide was enacted against our parents' and grandparents' generations, one to two generations ago. I personally have embodied the impacts of genocide within my father's lived experience. This genocide did not just lead to our displacement from our ancestral lands but is also why we have to constantly focus on healing in order to make it through in life. However, like banana trees, we become resilient while displaced and start growing our roots on the new lands that we now inhabit. Every time I think of the role plátanos played in my father's life, I think of how they played a role of survival. He recounts,

> Plátanos were a survival food during the war. That is all I would eat for months. That was the only food available to us in the guerrilla encampments. We had several plátanos in our area and whenever I eat one today, it reminds me of those times. I learned how to love plátanos and enjoy them even if it was the only thing I ate for months.

The irony in his story is that bananas ended up becoming a survival food for many during the civil war in El Salvador despite the large corporations that further commodified them also playing a major role in the civil wars. For example *Chiquita,* which was called the United Fruit Company at the time, in partnership with the Central Intelligence Agency (CIA), coorganized the coup that took place in Guatemala in 1954.[14] The political propaganda and aftermath of this coup fueled the Guatemalan civil war in the 1970s and eventually made its way to El Salvador.[15] The company played a major role in the civil unrest that led to the civil wars (genocides) across Central America. We, as Indigenous peoples of Central America, have a unique relationship with plátanos (bananas) because they have been embedded into our traditional foods like tamales but we recognize that those who were in power over these fruit plantations were in favor of the atrocities our communities had to face. It is also interesting how the term *banana republic* was created within the United States to refer to our Central American countries because our countries were deemed as politically unstable and our economy is heavily dependent on foreign economies and international fruit trade (e.g., banana plantations). During the violent capitol protests that occurred in the United States on January 6, 2021, several newspaper articles with headlines such as "Is the U.S. on the Verge of Becoming a Banana Republic?"[16] or "Is America Becoming a Banana Republic?"[17] started circulating. However, these articles failed to recognize the large role the United States played in creating the political turmoil that occurred in Central American countries and backing the civil wars such as the one in my beautiful El Salvador. The term *banana republic* was first introduced to the United States by short-story author O. Henry, who first used it in his story entitled "The Admiral," which read as follows:

> In the consultation of this small, maritime banana republic was a forgotten section that provided for the maintenance of a navy. This provision—with many other wiser ones—had lain inert since the establishment of the republic.[18]

The term *banana republic* is condescending and harmful to those of us who are the products of the US-backed civil wars in Central America. Seventy-five thousand of our people were murdered in El Salvador and two hundred thousand people were murdered in Guatemala during the civil wars.[19] The United States created political turmoil in our countries because they wanted to support the political leaders who were in favor of further privatizing the lands in our countries to sell to these foreign corporations. Unlike other ancestral foods, bananas, or plátanos, hold on to a recent history that continues to impact our people to this day. War trauma, as my father has explained, continues to impact those who had to fight in it.

> *How did war impact you?*
>
> Not very well. War is never something light to take. It mentally impacts us in a negative way . . . [long silence] . . . I healed over time from my years in war, but I was a child,

eleven years old. I can say that I still hold on to that trauma. War leaves us thinking that we will never be safe, even after you have left that place where you built those memories. I still do not feel safe. War instills this fear we carry on for years.

The capitol riot of 2021 in the United States does not compare to the political turmoil Central America underwent prior and during the civil wars.[20] The civil wars were not just political turmoil over elections but also genocides targeting Indigenous peoples and poor civilians who were tired of all the oppression they were facing every day. They were not just based on election results that did not favor those who stormed the capitol in 2021. Victims of the civil wars in Central America were being oppressed and underwent genocide. However, political turmoil is often associated with and compared to what happened in Central America despite it not being anything like what truly happened in these countries. For example, many children were forced to play a major role in the civil wars by either joining the opposition group (guerrilla) or the government's military. It took my father years to finally accept that he was just a child when he was forced to join the guerilla. Once he realized and accepted that he was indeed a child, he started healing and comforting that eleven-year-old who was scared and feared for his life while he endured the war for years. It was not until he was in the brink of life or death that he was forced to leave his country and journey through Guatemala and then Mexico. He recounts,

> I was eleven years old, but I did not consider myself young or a child. My father had passed years before the war started, so at an early age I had to start working to support my mother and four younger siblings. I taught myself how to catch fish so that I could go into the bigger towns and sell the fish. Given everything I had to do at such a young age, I never saw myself as a child, even once I joined the war. My childhood became

nonexistent, and I was forced to grow up in order to survive, right?

Yes, the legacies of bananas and the word itself have been integrated into our history. Being from a country that is often referred to as a banana republic demonstrates the complicated history and relationship we have with bananas. Not the fruit themselves but the tycoons and people behind these giant multimillion-dollar corporations. They tried to use bananas as a political, social, and economic tool against the people of Central America, but in the end, bananas have become an integral staple in our traditional foods. From our delicious *plátanos fritos* (fried plantains) to our tamales, they are now a part of our identity. For us, the younger generation, they have become a part of our ancestral foods that were introduced to our native lands as a commodity to control our countries politically and economically. This shows how even our ancestral foods adapt to our current landscapes and resources. However, to me, bananas also supported and nourished many of our people during war times. Like my father recounted, bananas became his survival food as the trees in their guerrilla encampment were the only source of food they had. Cooking with banana leaves makes me reflect on the history of my Indigenous community. From colonization to the civil wars, bananas were always there present in those historical events that still impact us today.

However, outside these large agricultural and extractive processes is an underlying threat that continues to impact Indigenous communities—climate change. Large agricultural corporations are responsible for the deforestation and contamination and are contributing to our accelerated climate change. Climate change is altering global temperatures and has resulted in an increase in natural disasters, including those that have severely impacted Central America, like hurricanes Eta and Iota in 2020. With natural disasters, the impacts of the Central American civil war also resurface, as it amplifies the vulnerability of Indigenous peoples and poor civilians.

Hurricanes in Central America

Hurricanes Eta and Iota devastated many Central American countries, like Guatemala, Honduras, El Salvador, Nicaragua, and Belize, from October to November 2020.[21] Many of our Indigenous communities in Central America live near or are in proximity to rivers, and with the heavy rains hurricanes Eta and Iota brought, the rivers were flooded. This flooding destroyed many homes and everything many families owned. The Indigenous communities who resided at higher altitudes experienced landslides that also destroyed their homes and caused everything that these landslides came in contact with to vanish. These landslides and other environmental impacts caused by hurricanes blocked many roads, making it difficult to bring any aid to the Indigenous communities most impacted. The government was always absent, and no aid was ever provided to our communities, so many people had to become resourceful and seek aid elsewhere.

Oftentimes when media outlets cover hurricanes, they focus more on home and shelter destruction but rarely mention the agricultural losses Indigenous communities face due to the natural disasters.[22] In Guatemala, many Indigenous communities lost their crops and their livestock. In El Salvador, many families in my *cantón* (pueblo) were trying to save their chickens, and some of these efforts were not successful. Not only do people have to seek refuge, but they also need to worry about their livestock and crops as this is sometimes their only source of food, and the loss causes many communities to face food insecurity. My family in El Salvador tried to get rid of the water that was flooding their homes, but it started raining so heavily that they had to go uphill to the highlands to escape the flooding. They had to carry elders on their backs as elders are among the most vulnerable during natural disasters. While the highlands gave them some refuge from the flooding, they were still at risk of landslides. Again and again we are witnessing how hurricanes continue to hit and impact Latin American and Carribbean nations the hardest. We also witness the lack of governmental aid and support provided to Indigenous communities during these times. My uncle Mario recounts his experience with these hurricanes and the impacts it left on El Salvador.

How did the hurricanes impact the community?

It really rained a lot that it ended up flooding the river that is near our homes. The river flooded so much the water dispersed and flooded some of our items. Families tried to save anything they could, but with water, things got destroyed easily. Our neighbors were trying to save their beds, even if it was just the mattress, so they had somewhere to sleep, but were unsuccessful. We have an elder who was trying to save his chickens, so we carried him and his chickens up to higher altitudes, while being careful as heavy rain can also create landslides. While we were waiting for the rain to stop and the river to stop flooding, we could see many of our homes flood. There is this sense of loss and anger that comes as you are watching the waters destroy everything you own. One cannot do anything to stop the waters or the river from flooding. We work hard to obtain the little that we have, and it is very hard to see how everything can be lost in the blink of an eye or storm. Our river had never flooded this severely. The waters seemed angry as the water was also flowing at rapid speeds within the river. It was as though Mother Earth was angry and it took a lot of trees as well in these rapid currents that developed in the river. The fishermen were all sent home when we received news of the hurricane. So while you witness how everything you own is lost, you also know that you do not have your source of income during this time.

In El Salvador, families were evacuated, especially those most at risk of landslides.[23] Artisanal fisheries were closed, which is something a lot of Indigenous communities rely on. Coastal areas are at higher risks of hurricanes, which is why most coastal Indigenous communities of Central America are impacted the most. Strong winds that destroy and bring down trees are also another leading cause of death during hurricanes. Also, the destruction of trees, in particular fruit trees that are found all throughout Central America, is another form of food insecurity our communities face during natural disasters. The devastation and destruction hurricanes leave are oftentimes compared to the devastation and destruction the war left on our communities, and this demonstrates how war trauma resurfaces for our community members during natural disasters. Many members of my paternal community came back to try to salvage some of their belongings, but they were all destroyed. My uncle recounted how many elders in our *cantón* were discussing how it was similar to reliving the war.

> Many elders were mentioning that seeing everything they own lost was similar to when the military would come to our *cantónes* and burn homes down. This is because the *escuadrones de la muerte* were not afraid to instill this fear and sense of loss to anyone, as their whole tactics were to calm the opposition groups through fear and loss. I think that because many elders, and even our generation, only faced such loss during the war, this is where our minds go when we witness natural disasters destroy everything. I was very young during the war, so I was not forced to fight in it like my older brother, your dad.

It is hard to witness what our elders and those who were old enough to remember the impacts of the civil war have to go through when they experience natural disasters because many of them associate loss to the war, and this is the magnitude of the trauma that war leaves on civilians and those who are forced to live through these human-violent atrocities. My uncle shares that this is because sometimes the results of natural disasters are the same as the results of war, which are not peaceful, and I can only imagine what their lived experiences in war were like. Through my father

I have been able to witness his healing but also his trauma, but it is just a glimpse of what he endured as, unless I can walk in his shoes, I will not know what demons he has to fight every day. This is how natural disasters resurface the trauma that is embedded in many of our Indigenous communities, and this is why not being supported or provided any aid during these times can make them fixate on their traumatic experiences only.

Hurricanes Eta and Iota left many homes under water, and our community members lost everything. All they cared about at that time was their lives, and they did not take anything with them while they tried to evacuate. These hurricanes had less of an impact on El Salvador than it did on neighboring countries like Guatemala and Honduras, but it still left some destruction. Given the topography of Central America, our landscapes are more prone to landslides that can cause damage.[24]

We witness how nature can serve as our sanctuary, even in times of natural disasters. Our people sought refuge in the hills from the flooding, and this reminds me of what my father shared about how he also sought refuge in nature during war.

> Nature had a protective role for me during the war. It also served as a refuge because I could climb the hills and hide under the tall grass during the war. When I wanted to escape my reality, I would climb the banana trees and get the most-ripe bananas while I was there. Nature served as a protective shield for me, and it continues to be that protective shield to our people.

Hurricanes Eta and Iota also energized the advocacy of the Indigenous communities of Guatemala that has been going on since 2004. The Maya Q'eqchi community has been advocating against the construction of the hydroelectric dams on their two rivers: Canlich and Cahabón.[25] In December 2015, the Guatemalan government granted construction permits to the Energy Resources Capital Corp. group to continue constructing hydroelectric dams along the river. They had built Renace I in 2004, the first hydroelectric dam, along the river and in October 2012 they built the second hydroelectric dam, Renace II, along this river as well.[26] In 2014, the company was granted approval to build Renace III, IV, and V. As of now there are a total of four Renace hydroelectric dams and plants on the Cahabón, Oxec, Canlich, and Chiacté rivers.[27] The Maya Q'eqchi communities fought against these construction permits, questioning their legality and validity, so these projects were delayed from when they were originally proposed and approved by the Guatemalan government.

Like most of our natural resources, the rivers are sacred to this Maya community; however, the government did not consult them when it authorized these permits. As a result, the leader of the resistance movement, Bernardo Caal Xol, organized his communities to oppose the construction of these dams and question the validity of the permits.[28] The construction of these dams ended up eliminating the water access twenty-nine thousand Maya people once had. Dams are human-made structures that impact the natural flow regimes of the water, sometimes blocking existing water passages in the river.[29] As my elders always told me, even our ancestors were well aware of the water flow of our natural water reservoirs like rivers, so they never built their homes near areas along the river that were more prone to flooding. Unfortunately, for many Indigenous communities like the Q'eqchi, corporations do not consult them and yet build dams that will alter the water flow, making some of their homes more prone to flooding. This is one way extractive corporations that govern or manage our natural resources end up dismissing ancestral knowledge and increase Indigenous communities' vulnerability.

For the Q'eqchi community, advocating for their rights and expressing their opposition to the building of the dams resulted in their leader being sentenced. Despite the constitutional court ruling in 2017 that Indigenous rights were violated, the hydroelectric plants and dams continued operating, including new constructions. The companies were ordered to consult with the communities, but these same companies had filed criminal charges against Bernardo Caal Xol.[30] He was accused of being violent toward several company and construction employees, and he was arrested. In November 2018 he was sentenced to seven years and four months in prison for these charges. Despite Amnesty International's involvement, he has not been released, and in 2020 his community, along with other Indigenous communities, experienced the hurricanes.

The role these hydroelectric dams and plants play with the recent hurricanes Eta and Iota in Guatemala is that the dams created diversions of the river, increasing and exacerbating the floods caused by the hurricanes. According to the Q'eqchi people, the companies that built these hydroelectric dams and plants are now controlling the river as they also chose to open the dams to avoid flooding and let this flooding be transported into the communities. They know that despite this, the Guatemalan government will not do anything against these dams and will continue to authorize their operations. What the Maya Q'eqchi community faced was not only the impacts of climate change but also the impacts of capitalistic corporations that are now the rightful owners of extractive methods that impact our natural resources, such as rivers, and create more chaos and harm during natural disasters.

We also gain a lens of how governments are complicit in how exploitative these companies are against Indigenous peoples. The government should release Bernardo Caal Xol from prison because he did not commit a crime. It is not a crime for Indigenous peoples to advocate and fight for our rights, but governments make it a crime. A sentence of more than seven years is also uncalled for. His imprisonment was to teach Indigenous peoples a lesson—a lesson that if they speak up, organize, and mobilize their communities, they will be imprisoned. The charges are false, but we continue to witness how racial hierarchies in Latin America continue to discredit and oppress Indigenous peoples. Not only do Indigenous communities have to adapt to human-made structures like dams, but they also have to adapt to climate change as climate change is already impacting their food sources.

Boiling Banana Leaves: Climate Change Impacts on Traditional Foods

We have to boil banana leaves and then set them to dry and cool. Once they have been boiled, dried, and cooled, they are ready to be used to wrap our tamales. The rising temperature of the water as it reaches its boiling point reminds me of how our Mother Earth is also dealing with rising temperatures that continue to impact our environment every day. It also reminds me of how climate change impacts are drastically decreasing our access to our traditional foods and how this leads to food insecurity in many of our communities. The swirling of the boiling water also symbolizes the hurricanes that have drastically impacted Central America during my lifetime. My cousin Dalia, who oversees the communal milpas of our pueblo, has described how climate change has impacted agriculture in Oaxaca:

> It has affected the crops severely. Many crops have not grown correctly. For example, in the case of corn, it has affected its growth because that it does not rain when it should or it rains heavily destroying most of the milpas, systems where corn is grown.

However, aside from the impacts climate change has had on our land resources, it has also severely impacted our fisheries and marine resources. In San Mateo del Mar resides the small coastal pueblo of the Ikoots people. They rely heavily on fishing as this is an important and essential component of their overall livelihoods and well-being. However, with climate change, they have noticed a decrease in fish stocks and catch. This, coupled with the COVID-19 pandemic that began in 2020, has made this community face food insecurity. Climate change is impacting fisheries globally and this ultimately impacts Indigenous communities that rely heavily on fisheries. Climate change is causing water temperatures to rise,[31] and while the change in temperatures may not impact humans as much, it severely impacts aquatic animals and ecosystems. The rise in temperature results in a huge decrease of fish stocks or populations. It also decreases the pH of oceans, making them more acidic. This is also known as ocean acidification, which destroys habitats, severely altering marine ecosystems.[32] Many aquatic organisms have shells and skeletons that are made of calcium carbonate, which dissolves in acidic conditions. The acidic conditions impact fish's ability to grow and severely impact their neurological systems. Other aquatic organisms that are severely impacted include coral reefs, crabs, urchins, oysters, and shellfish, many of which Indigenous coastal communities also rely on.

In marine conservation, MPAs have been established to combat the impacts of climate change.[33] These MPAs are aquatic wildlife sanctuaries that aim to protect endangered aquatic species by preventing any fishing or recreational activities from taking place in the designated MPA. However, Indigenous peoples globally continue to advocate for their rights and voices to be included in the development of MPAs as not being able to fish severely impacts their livelihoods and well-being. In Latin America, an estimated 8 percent of its oceanic territories are MPAs.[34] With proper consultation and involvement of local Indigenous peoples, MPAs can serve a positive purpose.

However, time after time, we are reminded that as Indigenous peoples we have to face the highest ecological debt despite not being responsible for climate change. Climate change is a result of Western extractive methods introduced on our lands under the premises of capitalism. The way we became heavily dependent on energy sources like fossil fuels and oil has created a large ecological debt that burdens Indigenous peoples the most. In the case of our fisheries, the overfishing introduced into our waters created these boom-bust cycles[35] that rapidly decreased our fish stock because they were overharvested and fished by settlers and new-comers to our lands. Commercial fishing and the seafood industry continue to profit from selling and importing fish while not supporting the same Indigenous peoples who now have to compete for these fish stocks. Oil refineries that are also near the coast of San Mateo del Mar are also impacting their fisheries as all the pollution produced from these refineries that are only twenty-eight kilometers from their communities end up in their coastal waters. Ana Laura Palacios Cepeda shared how climate change and the local refineries are impacting her community's fisheries.

What are the traditional plates of the Ikoots community?

The traditional dishes are fish tamales, shrimp mole, grilled fish, *capeado*, black mole, beef stew, and different seafood stews, bean tamales, and barbacoa, stingray salpicón.

How is climate change impacting the fisheries?

Definitely, climate change has impacted our fisheries. Times have changed and every day there is less access to fish. It worries our fishermen, but sometimes there is not anything we can do. Adding to this, the refinery near our community continues to impact us. We see all this pollution run down to our coast, and there is nothing we can do as the refinery is not going anywhere.

What are the Ikoots baskets you started through Manos del Mar Oaxaca?

An Ikoots basket is a small basic basket with local products from San Mateo del Mar. The basic basket has *totopos*, fish, shrimp, flour, bread, coffee, to mention a few things. In this way, we promote the local economy and at the same time get to support a family in San Mateo del Mar who is currently in need.

Her story demonstrates that Indigenous peoples are resilient and also productive. Despite her community facing food insecurity, she has created a mutual aid project, *Canastas Ikoots,* that translates to "Ikoots baskets." As she mentions, these are small baskets that provide essentials and traditional foods to families in need in her community. Oftentimes we read stories that highlight our Indigenous communities' vulnerability, but these stories end up harming us more because they create this cycle of *white saviorism* that makes people assume we need someone to save us. As Bryan van Hulst Miranda analyzed and described, white saviorism roots itself in "Brown victimhood that urges European descendants to take up the moral duty to save the oppressed from their oppression. It is founded on a white egocentric myth that there is something that Brown people must need that only white people can give them."[36] Indigenous peoples do not need saving. My elders have always reminded me that due to our strength, we have been able to survive all these years despite the impacts colonization left in our communities. Yes, it is sometimes not an easy life, as my elders would say, but it is a life filled with happiness, strength, and resilience. We do not need anyone to save us. What we need is people to take responsibility for the actions and oppression that continue to be enacted against us. Our traditional foods continue to be jeopardized, yet it takes our own people to work hard to address the food insecurity we face as a result of climate change, coupled with human impacts such as the oil refineries that are polluting the Ikoots community fisheries.

White saviorism dismisses our autonomy and resilience as we have learned that the government does not have the best interest for our communities. This is why we become resourceful in finding ways to help our families and communities. In the case of Ana, she did not wait for the government to allocate any aid to her community, which the Mexican government has not done to this day, but instead found a way to create a mutual aid project to continue supporting her community. We need to move away from this white saviorism as it is a modern-day version of *manifest destiny* that determines that it is up to settlers to continue saving us. What we demand as Indigenous peoples is for our rights to be protected, for conservation agencies and corporations to consult us, and for us to regain the autonomy to manage and oversee our natural resources. Climate change continues to heavily impact our communities, and in the era of climate change, Indigenous peoples need to be placed at the forefront. Unfortunately, many coastal communities are facing severe impacts on their fish stocks, such as the Coast Salish communities of the Pacific Northwest where I currently reside.

Climate Change and Salmon in the Pacific Northwest

The Pacific Northwest has a long history of Indigenous advocacy and resistance for food justice, in particular salmon. In order to understand how tribal rights differ from Indigenous communities of Latin America, we must understand the role treaties play. Treaties are federal agreements that were signed between the of the United States federal government and tribes in order to protect their rights within the new federal policy frameworks.[37] The Coast Salish tribes and nations signed the Stevens Treaties between 1854 and 1856. In these treaties there was a fishing clause that granted the tribes the right to access and harvest 50 percent of fish stocks in their accustomed areas.[38] For instance, the Treaty of Point Elliot's fishing clause stated,

> ARTICLE 5. The right of taking fish at usual and accustomed grounds and stations is further secured to said Indians in common with all citizens of the Territory, and of erecting temporary houses for the purpose of curing, together with the privilege of hunting and gathering roots and berries on open and unclaimed lands. Provided, however, that they shall not take shellfish from any beds staked or cultivated by citizens.[39]

These fishing clauses were incorporated in all treaties but were not fulfilled by the federal governments, resulting in an environmental injustice against the tribes that took years to be addressed. The treaties were dictated by President Thomas Jefferson to try to assimilate Native Americans by introducing them to agriculture (e.g., farming, raising cattle, etc.) and abandon their old ways of hunting, gathering, and fishing.[40] Given the importance of salmon to the Coast Salish peoples, the tribes initiated the civil rights movement of the Pacific Northwest in the 1960s as an act of resistance and resilience against the treaties not being honored and their fishing rights not granted. This is also known as the Fish Wars, and some of the key actors in this grassroots movement included Billy Frank Jr.[41] They held fish-in protests, which were based on the sit-ins that took place during the civil rights movement in the American South.[42] They were arrested because the participants were fishing without federal licenses or permits from the Fish and Wildlife Services. They were refusing to obtain state licenses as a form of self-determination and to reassert their treaty rights that were never honored, but were advocated for by their ancestors through the treaty signing processes. The fish-ins finally garnered media attention in 1964 and prompted celebrities to join the movement, including Marlon Brando.[43]

After countless battles at Frank's Landing—the location where the fish-ins mostly took place—the National Association for the Advancement of Colored People (NAACP) began providing legal support to the tribes. The NAACP fundraised for the legal fees and with their help, along with that of the American Civil Liberties Union, and ultimately, the US Department of Justice, this case was taken to court. In 1974, a court decision made by Judge George Boldt changed the course of history for the Pacific Northwest and Washington tribes.[44] This court ruling became known as the Boldt Decision, and it asserted the treaty rights the tribes were denied for years. One of those rights was the fishing clause that finally allowed tribes to practice their fishing rights as a sovereign nation without needing to apply for fishing permits from the Washington Department of Fish and Wildlife. The tribes now had the right to 50 percent of the fish stocks in the state of Washington.[45]

Despite this amazing victory led by the tribes of the state of Washington, salmon are endangered due to the habitat loss, coupled with the environmental impacts that are occurring in the state of Washington. This reduces the number of salmon the treaty tribes can catch for sustenance and cultural identity. In 2020 to 2021 the treaty tribal quota for catching salmon was established as 35,000 Chinook salmon and 16,500 Coho salmon.[46] There are twenty treaty tribes in the state of Washington that signed treaties that are laws and regulations the state has to abide by, and the tribes have to abide by these numbers. However, for recreational fishing, the quota was established as 26,360 Chinook salmon and 26,500 Coho salmon.[47]

It is important to note that for the Washington state tribes, salmon is their cultural and traditional food that plays an important role in their creation story. This means that salmon is also their relative and essential to maintaining their cultures. However, because of the boom and bust that the fisheries in the Pacific Northwest experienced when salmon populations were thriving there, coupled with the climate changes and habitat destruction, salmon populations have drastically decreased. Unfortunately as of 2021, salmon have been reported to be on the brink of extinction. Our oceans and other bodies of water continue to increase in temperature, which kills many of the adult salmon during their migration route. In 2019, Washington state allocated $18 million to fund salmon recovery projects.[48] However, reports in 2021 state that there was a need for $4 billion in order to fund the important projects needed to restore habitats that continue to be destroyed.[49] Now the state of Washington tribes are also facing the risk of losing the important cultural component that salmon is. Salmon is also medicine, tradition, and ancestral knowledge that the Washington state tribes continue to sustain to this day. Washington state tribes have served as comanagers but when it comes to who has more to lose, between the two comanagers—the state of Washington and the tribes—the tribes have more ecological debt. Saving the salmon saves the Indigenous peoples whose cultural identity is parallel to salmon.

Protecting salmon has been one of the environmental justice movements the tribes in the state of Washington have continued to lead in this state. However, the situation for salmon continues to worsen because of climate change. Unfortunately, we continue to witness the impacts climate change has on Indigenous traditional foods and, despite mitigation and adaptation efforts, not much is done to support Indigenous communities during times of need.

Preparing the Tamales with Banana Leaves

The last step is to prepare the tamales inside the banana leaves. Everything that goes inside the tamales is placed on top of the *masa,* dough, and is then wrapped in the banana leaves. This last step makes me reminisce on the importance of our traditional foods and the role they play within our spirituality. Every year we celebrate our spirits in a tradition known as *Día de los Muertos,* Day of the Dead. The ceremonies we hold during this tradition are filled with our traditional foods, such as our tamales. We place tamales and other foods on the altars we make honoring our loved ones who had journeyed to the spiritual world. This is how we greet them when the doors from the spiritual world open to the world of the living. Our traditional foods even play a vital role during the traditions that surround funerals. Our traditional foods embody our spirituality, which is important to our identity and existence, even in the diaspora. This is why food sovereignty and justice have been movements that continue to build among our communities.

When we face food insecurity, we support one another through the Indigenous principle of reciprocity that is deeply embedded in mutual aid. Mutual aid has become an integral and important component of our food sovereignty. The exchanges allow other community members to support one another in times of need. In San Mateo del Mar, Ana, along with her collective, Manos del Mar, has led mutual aid efforts to be able to provide the Ikoots baskets to her community. In San Pablo Tijaltepec, Lita, along with her collective, Ñaa Ñanga Tijaltepec, has led mutual aid efforts to provide weekly groceries to families severely impacted by the loss of their communal harvest. Mutual aid is how we have organized within our pueblos as when one family is in need, other families come together to try to support that family. This is one of the Indigenous principles that has survived colonization as settler colonialism favors individualism over community. My elders always told me that when I had an extra pair of shoes, instead of putting one pair away, see who needs that pair and give it away. Having this community support, and not only in times of need, has been one of the reasons why our Indigenous communities have been able to thrive during times of high food insecurity. We know firsthand that the government will never provide any aid, and this is why we have never depended on the government when we needed something. Mutual aid is grounded in our Indigenous teaching of supporting one another, especially during hard times. We know that like milpas, the care and dedication we provide to our community will always be reciprocated.

Wrapping the tamales in banana leaves serves as the metaphor that despite it all—climate change and political turmoil that continue to oppress us Indigenous peoples—our resilience is as strong as a tamal once it is wrapped with banana leaves. Our existence and history as Indigenous peoples are intertwined like the veins in a banana leaf. The same way we come together to enjoy tamales, we come together to support one another in times of need, and this needs to be celebrated, honored, and supported by allies and non-Indigenous peoples.

8

Indigenizing Conservation:
Healing Indigenous Landscapes

I live my life embodying the teaching my grandmother instilled in me—
that no matter which lands I walked on, I had to learn how to build
relationships with the land and the Indigenous peoples whose land I
reside on to become a welcomed guest. As a displaced Indigenous woman,
my longing to return to my ancestral homelands will always be there,
and this is why I continue to support my communities in the diaspora.
However, my relationships are not only with my community, but also the
Indigenous communities whose lands I am displaced on, and this is the
foundation of my work while residing in the Pacific Northwest. I strongly
believe that in order to start healing Indigenous landscapes, everyone
must understand their positionality as either settlers, unwanted guests,
or welcomed guests, and that is ultimately determined by the Indigenous
communities whose land you currently reside on or occupy.

This teaching has also helped me envision my goals in life. Every day I
get closer to becoming an ancestor because life is not guaranteed but rather
a gift that we are granted from our ancestors who are now in the spiritual
world. I think this is the premises of healing Indigenous landscapes, as

once we understand our positionality, we can take actions to dismantle the systems that either oppress us or benefit us. The systems embedded within settler colonialism continue to benefit settlers, and settlers are defined differently in each country. Even Latin American countries operate in different settler colonialism frameworks, but these ultimately favor all settlers in each given country. This is why I say that many Latinos play the role of the oppressed and oppressor. Being Latino does not determine that one is not a settler back in Latin America, as the racial caste systems that operate in Mexico and throughout the Americas favor whiteness and those who are closer to the settler identity than it does Indigenous peoples.

Another teaching that comes to mind is how we, as Indigenous peoples, are taught to share our sacred teachings, but they end up being co-opted or appropriated. I believe that it is important to understand our positionality and that while some Indigenous peoples have shared their teachings and knowledge with non-Indigenous peoples, non-Indigenous peoples need to stop taking ownership over it. I think this is a big problem in academia, as there is a lot of Indigenous scholarship written by non-Indigenous scholars or scientists. We need to stop reframing this narrative that Indigenous peoples need their stories told *for* them, instead having their stories told *by* them. As Indigenous peoples, we know that our cultures have been tailored to be consumed by a non-Indigenous audience, especially in the environmental sciences. The romanticization and fetishization of Indigenous cultures continues in the twenty-first century and has worsened over the years due to social media. While social media has served as a positive tool to us, it has also increased the romanticization of our cultures and the theft of our knowledge systems. This is why I have learned to be careful of what I share online, as I have witnessed too many times the theft of Indigenous peoples' knowledge by non-Indigenous peoples who take what they have learned and proceed to publish it in news articles, on blogs, and even in research peer-reviewed journal articles. We also see how some social media accounts promote living on Indigenous sacred lands while further displacing Indigenous peoples. This is mostly done through the promotion of tourism as well, and while tourism can benefit a country's economy, it continues to jeopardize the Indigenous landscapes and peoples.[1] People want to consume our Indigenous cultures, but without the Indigenous peoples, and this objectifies our culture for the white gaze, making it palatable to white people only. Yes, there are Indigenous peoples who uphold stereotypes, but this is because they are tailoring everything to the white gaze, and oftentimes not to Indigenous peoples. People need to stop objectifying Indigenous peoples as people from the past, through a historical lens, and acknowledge that we have adapted to modern times as well. We continue to pass our teachings to the younger generation to try to not just heal ourselves, but heal our communities and landscapes. Our landscapes are where our creation stories take place, where we continue to practice our culture, and where we continue to fight against the eradication and assimilation tactics settler governments continue to practice. Healing our Indigenous landscapes will allow us to heal ourselves, our people, and break free from the chains that settler colonialism continues to restrain us with. This includes being able to go against conservation practices that further oppress our communities and being able to say no to tourism ventures that will harm our communities further.

Colonization Metaphor through Removing Invasive Species

As mentioned before in this book, there is no word for conservation in many of our Native and Indigenous languages. While there are some phrases close to what conservation means in Zapotec, most of these words relate more to "taking care of" or "looking after," which is not truly embodying what conservation means.[2] When healing landscapes, the word that is used to do this is coined as *restoration*. Restoration teaches us that in order to heal a landscape, we must get rid of all the invasive species that are known as *weeds*. However, this fails to truly heal the entire landscape as it only focuses on one component, invasive species, and not on other factors that might be impacting the entire ecosystem or landscape.[3] I have sat in many presentations about invasive species where they have been called the devil, evil, or nightmares. However, the irony that lies within these descriptors is that for many who practice restoration or are in the environmental sciences, most of these invasive species are their plant relatives as these were introduced during colonial times by settlers and colonizers. What this means is that many white people have lost their ancestral roots due to the assimilation the Americas have undergone and, as a result, they have lost their relationships with the same plants they now deem as terrible beings. Yes, invasive species harm an entire ecosystem, sometimes outcompeting all native plants in this same landscape; however, we are taught as Indigenous peoples that regardless of whether this plant belongs there or not, we must ask its spirit for permission. As I shared before, we acknowledge them as displaced relatives rather than invasive species, since at the end of the day, they are also someone's plant relatives. What Western conservation, environmental sciences, and restoration continue to teach us is that anything that is not native is not welcomed to the flora or fauna landscapes. However, this rhetoric is never applied to humans as we seem to be the exception for our own laws, rules, and regulations that we only apply to our environments. This alienation is only applied to vulnerable communities such as our Central American climate and war refugees because they are ostracized thanks to current immigration laws.

Removing invasive species without good intent or connecting with them causes scars. When I was taught restoration practices in my academic courses, I was taught to work hard and fast to complete the task. In my courses, relationship building and asking for permission were never mentioned when we were instructed to remove the invasive plant species or weeds. Being the only Indigenous person in many spaces, we sometimes opt not to speak up or mention anything as sometimes we are questioned, ridiculed, and labeled as ignorant. Yes, most of our practices do not make sense under the Western science lens, but we should not have to alter or adapt our knowledge systems to fit the Western science lens. Our Indigenous knowledge and practices should be acknowledged. I recall the many times I was ridiculed by white teachers and professors, and this instilled in me some form of shame that took years to heal from. This happens a lot in academic spaces as we are deemed to be ignorant, naive, and inferior, and many continue to hold to these unconscious biases that end up harming not only their Indigenous students but also their Indigenous colleagues and people outside of the academic realm.

My experiences as an Indigenous student in the environmental sciences have shaped how I teach and navigate my own courses. That young Indigenous woman who was sometimes ashamed to share her teachings or knowledge is now leading and teaching such courses, so I ensure I center my own teachings. This makes a difference to Indigenous students as many of them have come up to me and told me that my class was a space where they did not just feel welcomed but also acknowledged. While they may be quiet and timid in other courses, they are eager to share their knowledge and cultures out loud in my courses. I wish they felt this sense of belonging in all their courses, but given the few Indigenous faculty members across universities, especially in the sciences,[4] there is a long way we have to go as a nation that continues to have educational disparities.

I recall when I taught my first restoration class, one of my students pointed out the language I used and how this was different to him because professors tend to use academic jargon and terminology that is not accessible to those not in academia. He deeply appreciated my use of nontypical language because he did not have a Western science background and felt more at ease with the language I was using. Yes, I would use words and phrases like, *friends, they do not like each other,* or *displaced relatives* when I referred to the plants (flora) of the landscape we were restoring, then go on to explain to students what the equivalent of these relationships was in Western science. For example, the phrases *they do not like each other* or *friends* refer to the competitive or mutual relationships plants can exhibit with each other. These relationships are identified through plant guilds. Plant guilds allow us to find out which plants can coexist and thrive in the same community as some might outcompete other plants for nourishment or even sunlight.

For me, healing Indigenous landscapes means centering non-Western ways of thinking, learning, and teaching. I can give a long presentation on plant relationships using scientific terminology, but it is best to frame it through a discourse that everyone can understand, and that includes my parents, who do not have an extensive Western education. My mother comes from a family of nine siblings, so she was only able to make it to the sixth grade, and my father did not have the opportunity to get any education as at a young age he lost his father, had to work, and then survive the war. I always tell myself that if my parents cannot understand what I am doing in my scientific work, I am not just failing them but also my entire communities, as educational opportunities continue to be granted to them.

By integrating not only nonacademic terminology but also hands-on projects in my restoration course, I was able to offer students a metaphor that explains colonization and the impacts it has on Indigenous peoples. After completing their restoration service hours, my students would complain about the cuts the invasive species would sometimes leave on their arms and legs. I would tell them that after one or two cuts, they would get used to it. However, we were removing Himalayan blackberry (*Rubus armeniacus*), and these are known for their long thorns that can penetrate about anything, even the protective gear we had on. Yes, they were hard to remove, and the cuts they would leave would hurt. Since I was doing the restoration work almost every day, the cuts would make it hard for me to wash my hands as it stung with soap or even just water. Therefore, I did understand what they were referring to; however, the cuts the Himalayan blackberry would leave reminded me of using this as a metaphor to teach my non-Indigenous students about the pain colonization has left on Indigenous peoples and communities.

The metaphor related to the pain these cuts would leave and how they symbolized components of the pain we Indigenous peoples continue to endure because of settler colonialism. This was also a percentage of that pain we carry as Indigenous peoples, because colonization has hurt us, fractured our communities, and continues to impact our Indigenous landscapes. Many of my students were non-Indigenous so using this metaphor allowed them to metaphorically grasp the pain. We were working in an urban space that was reclaimed by Indigenous peoples in Seattle and this allowed them to understand metaphorically the sacrifices that were made for this space to be reclaimed within an urban park. They witnessed how the twenty acres of land that was leased to the urban Indigenous organization that oversees Daybreak Star Indian Cultural Center had not been restored like other parts of the park. This is a 534-acre urban park and the walkways for tourists and pedestrians were cleared and maintained. However, once you walked into the jurisdiction of Daybreak Star Indian Cultural Center, there was no restoration work that was taking place or had been done there by Seattle Parks and Recreation. This meant that we were removing wild blackberries that towered over my five-foot stature—invasive wild blackberries whose roots were very thick and deeply embedded on the ground. Yes, my arm experienced a lot of pain and soreness as I led overall ten different groups of students in this restoration project. But that pain still does not resemble the pain that I carry as an Indigenous woman who is trying her best to continue uplifting her communities within the environmental discourse.

Restoration work is physically exhausting. However, it allows me to connect to the landscapes that are foreign to me as a displaced Indigenous woman. I strongly believe that we must build relationships with the Indigenous peoples whose land we occupy as well as the lands themselves. This means that we must provide our services and build these relationships through actions that support them both. I navigate new foreign landscapes knowing that they carry someone's animal and plant relatives, and these places are where someone's ancestors and spiritual guides continue to navigate. I reflect on the impacts the Indigenous peoples from these lands are facing. In my new environment in Seattle, I think of how the Duwamish people, whose lands this city was built on and who continue to reside here, have not even received federal recognition and are not consulted on city planning initiatives, policies, and management practices. Settlers must learn their own history and the role their ancestors played in this history, and also the Indigenous history that brings to light the atrocities, genocide, and violence that were enacted on the Indigenous peoples of these lands.

As an Indigenous woman of the Americas, I carry the history of the pain my ancestors had to endure, and in order to heal our landscapes we must heal ourselves as well. Everything that impacts us ends up impacting our environments as we are not separate from nature. We are a part of nature, and what impacts us impacts our nature and vice versa. Our Indigeneities are attached to this relationship with nature. Healing our landscapes ultimately means that land should be returned to Indigenous peoples and that we need to start calling out the colonial legacies that sometimes tourism advocates for. Tourism continues to further displace Indigenous peoples from their ancestral lands while also leading to environmental impacts and degradation.

Tourism: Its Colonial Legacy

Tourism is known to redefine landscapes as they become heavily transited locations and spaces that center economic revenue over anything else.[5] For example, tourism in Mexico is a very important and crucial industry for the country. For El Salvador, tourism contributes a large portion to the country's revenue and economy. However, we cannot forget that these same landscapes have an Indigenous history that is removed or utilized for commercial purposes. Thus, tourism embodies a colonial legacy as it continues to ignore, dismiss, and eliminate Indigenous peoples' relationships, histories, and legacies within these spaces. Unfortunately, many of the locations where we have our sacred sites, that are now referred to as ancient ruins, are now heavily touristic places—for example, the Yucatán Peninsula that extends over three countries: Mexico, Guatemala, and Belize. In Mexico, the Yucatán Peninsula extends through the states of Campeche, Yucatán, and Quintana Roo. The beaches and seascapes found in this region have made it a tourist attraction that is sought after by many. Cancun, for instance, is one of the most famous vacation and beach destinations due to its beauty and the region's rich Indigenous history as the epicenter of the Maya civilizations. However, as noted before, the Maya civilizations of the past are what are preserved by tourism and not the Maya descendants who are still fighting for their rights, livelihoods, and to protect Mother Earth. In the Yucatán Peninsula we have Indigenous Maya communities like the Yucatec, Mopan, Huastec, and Q'eqchi', among others, who are constantly fighting the government to prevent more sacred sites from being destroyed or built upon for touristic purposes. Two of the common land disputes that are being fought by the Maya communities include the *Tren Maya* (Mayan train) and the *Mirador* ecotouristic resort.

El Tren Maya

El Tren Maya (Mayan train)[6] is a train system that is being developed under the Mexican government to provide a method of transportation for tourists along the fifteen-thousand-kilometer coastal area where the popular beaches are located.[7] This train will go through all the popular beaches, continuing to displace the Maya pueblos and small businesses, many of them owned by Maya people, in the area. Due to tourism, many Maya people from the Yucatán Peninsula have been forced to work in this industry because this is the only economic revenue or opportunity they are granted. However, many of them have to take jobs such as performing or becoming a tourist attraction under the pretense of being frozen in time. The Yucatán Peninsula is known for its "ancient Mayan rituals," which depict Maya pueblos and people as left in the past and not as current and present people. Many Indigenous Maya people are also street vendors who sell their *artesanias*, handmade goods, and often face a lot of harassment by business owners or managers of elite corporations and name-brand boutiques found in the touristic area. On top of that, police corruption forces many of them to pay officers a share of their profits; otherwise, they are threatened with imprisonment. Police corruption is often more visible in tourist areas and, oftentimes, Indigenous street vendors face this the most. The creation of this train will desecrate many sacred sites, but this is not a priority to Mexico's current president, Andrés Manuel López Obrador.

His proposal of the train has been promoted to the Indigenous pueblos as something that will benefit them. In some of the memorandums that the president's office released, he promises this project will solve some of the infrastructure problems many Indigenous pueblos face, such as water access and irrigation. He has also described this as a way to solve some economic problems as this train will give them access to sell goods to tourists in other locations such as the train stations. The train project has proposed thirty stops and a creation of twelve train stations. However, it is important to mention that the creation of such a project will just mean that private investors will be interested in purchasing more land and further displacing Indigenous communities of the area. The Yucatán Peninsula is known to house privately owned luxury hotels, such as the Hilton hotels, among others. The government sells land without consulting Indigenous peoples, which further displaces them. The harsh realities Indigenous peoples had to face in the past are one of the reasons they continue to not trust the government. Also, there is a president elected every six years, *sexenio,* so while this current president might promise them something, the next president might not abide by those promises. In our communities we have a saying that as Indigenous peoples we get to meet every Mexican president, but while the Mexican government continues to say their intentions are to support our Indigenous communities, we know that they just do it for a photo op. Every Mexican president has at one point met with Indigenous communities and taken a lot of photos and has been followed by the media. However, we continue to be props, and nothing has come from such relationships or meetings. Some might call us cynical, but we are simply realistic to the effect settler governments have on our people. We continue to be their afterthoughts; that is why we continue to fight these government-led projects that do not consult us in the first place.

Building a train that will extend over 1,500 kilometers will result in a lot of deforestation. We already have many animal and plant species that are endangered in this region. Destroying their habitats and ecosystems will continue to decrease their populations and put them at higher risk of extinction. The Zapatistas (mentioned in chapter 6) have already opposed the president's proposition.[8] Some of their pueblos will be impacted by the creation of this train, and for them, the preservation of their sacred sites is more important than any capitalist venture. Many Maya pueblos have witnessed firsthand the desecration of their sacred sites, such as their *cenotes*. Cenotes are underground water openings that connect to underground caves that lead to the ocean. For many Maya communities and pueblos, cenotes (D'zonot or Ts'onot) are known to be the gates to the underworld, and they were the location of their sacred burials and ceremonies, making them important to their spirituality.[9] However, due to the tourism industry in the Yucatán Peninsula, these sacred sites are now open to the public and they can swim in them as a leisure activity. This has prevented the Maya communities from practicing their ceremonies and rituals and many of these customs have been lost as a result.

The spirit of the Sukan was a serpent-like deity that took care of the cenotes. These cenotes also offered them fresh water, but after the displacement of many of these communities, the Maya pueblos who once practiced their ceremonies and rituals in cenotes are no longer living in proximity to them or near enough to maintain this tradition. With all the toxins and chemicals released from tourism, such as the use of sunscreen and perfumes, the water quality of cenotes has decreased and continues to be threatened. However, since they continue to be important to the tourists' experience, no one enforces the cenotes' protection. This leaves Indigenous pueblos skeptical of the positive outcomes that tourism can bring to our communities. Thus, the Mayan train is not as accepted as the president and the Mexican government would like it to be.

We continue to experience this fascination with Maya sacred sites. As mentioned before, the Yucatán Peninsula extends to Guatemala and Belize, and in Guatemala, Maya sacred sites are also experiencing threats due to tourism purposes and ventures proposed by non-Indigenous peoples.

El Mirador and Hierve el Agua

El Mirador is located in El Petén, Guatemala, and is home to one of the largest Maya pyramids.[10] In 1990 the region became a conservation protected reserve, and in 2002 the Guatemalan government made it a designated area, which limited access to people outside of the region's perimeters. There was a lot of illegal logging, agriculture expansion, a spate of forest fires, and even drug trafficking that severely impacted this area and continues to do so. Protecting El Mirador has now taken over a decade of constant fighting against private businesses who want to completely control the area. The region is home to many emblematic species, one of them being the jaguar, *Panthera onca*. Given the importance of place-based history and, in this case, place-based animals, the Maya communities want to protect this area as it embodies their important history.

However, this is all under constant attack and threat due to an archaeologist, Dr. Richard Hansen. He built his career on the excavation of sacred Maya artifacts in this area and tied the history of this place to the Mormon history. The belief he supported through his excavations was that the Mormons' ancestors helped the Maya communities build the pyramids. As a result, he wants this place to be converted into an ecotourism destination to bring more Mormons from the United States to this protected area. He wants this area to become a resort and build hotels and other touristic attractions to bring in more economic revenue and other "opportunities" to the Maya communities of El Mirador and ultimately "save" this area.[11] He is also proposing building another train in this area that will help transport people to the reserve and ecological hostels. The United States and Senator Jim Inhofe, an Oklahoma senator, support this proposal with a bill backing a $60-million project on the oldest sacred site of our Maya civilization and history.[12] Hansen's attempts to take ownership of our Maya sacred site has led him to calling the Maya people and communities who oppose his proposal people "with no vision" as he proudly announced in an interview.[13] He goes on to make "smoking signals" to ridicule some of the sacred traditions that he was probably invited to by Maya communities for his archaeological work in ruins that do not belong to him or his history.[14]

Right now the conservation framework operating in this region allows the communities who heavily rely on the forest to protect the forest from illegal loggers, drug traffickers, and other impacts mentioned before. Hansen's proposal will convert this site into a tourist attraction and we already know that the theft and illegal trafficking of ancient artifacts that belong to our Maya people are a multimillion-dollar illegal activity that continues to impact these sites and locations.

Healing our landscapes means that we allow the local people to use their place-based knowledge to make the best decisions they deem will protect and heal their landscapes. However, we continue to see Indigenous people and communities portrayed as ignorant for "lacking a vision" as people like Richard Hansen have decided about us because we simply know that tourism will further destroy our landscapes. In Guatemala we continue to witness the missionary religious trips that continue to impact our communities but the Guatemalan government allows. Western religions, like Christianity and Catholicism, have now become important in many Indigenous communities, but we cannot forget that this was the same religion that led to the inquisition of our ancestors through manifest destiny.

This ongoing desire to "solve our ancient histories," such as how our ancestors built the pyramids continues to lead archaeologists such as Richard Hansen to our sacred sites. However, it is important to remember that it was white settlers, many of whom are the ancestors to these archaeologists, who ensured our people and civilizations collapsed with the introduction of disease and colonialism to our lands. The same people whose ancestors ensured we lost this knowledge among our communities now want to discover our Indigenous knowledge systems to further co-opt them as has happened in the past and continues to take place. Indigenous peoples, including those in the diaspora, have expressed their opposition against the tourism project, including a group I am a member of, *Mayab' Scholars in Diaspora*. Here is part of a letter written to Hansen, led by our leaders of this group:

> We are from diverse Maya Nations, defenders of life, deeply respectful of Mother Earth, defenders of our ancestral-contemporary territories, and of our millennial history and memory. As activists, weavers, writers, artists, and academics that live in and outside of Guatemala, we express our deep concern over the proposed Mirador-Calakmul Basin Maya Security & Conservation Partnership Act. Our nations were never consulted nor informed about this proposal and have never given their consent. We have seen the short news video from VICE News "Mayan Ruins in Guatemala Could Become a U.S.-Funded Tourist Attraction" (June 17, 2020), in which it informs us that you are leading a campaign in Guatemala and the United States to generate funding to transform the sacred El Mirador site into a tourist attraction and in that way bring "economic benefits to Guatemalans."[15]

What is happening with el Tren Maya and El Mirador demonstrates the importance of asking, consulting, and prioritizing Indigenous peoples and their needs, especially those whose lands and communities are severely impacted by tourism. For example, in San Lorenzo Albarradas, Oaxaca, the Mixe Indigenous community worked hard with other poor community members to close their territories to tourism because it was further harming them as opposed to it doing what they were promised tourism would do. The communities in San Lorenzo Albarradas have been struggling with their lands as they have been extremely demoralized that agricultural attempts to grow crops have failed.[16] Thus the local communities and landowners have been advocating for the closure of the tourist site for over eighteen years.[17]

Hierve el Agua is known for its beautiful waterfalls and topography; however, all the revenue was being made by tourist companies that were not supporting the local communities, including the Mixe community from the impacts tourism was leaving on their lands. In 2017 the Indigenous community brought their case to the *Sala Indígena del Tribunal Superior de Justicia,* which was created to oversee legal cases brought by Indigenous communities of Mexico that show violation to their land and human rights.[18] This tourist site is estimated to generate over US$96,000, but none of the funds have been distributed to the local communities.[19] It took years of advocacy from the local Indigenous communities, in solidarity with other local communities, to make sure their voices were heard.

This shows us that oftentimes, while tourism is framed as a great opportunity for local Indigenous communities, it really is not great at all. The privatization of tourism benefits companies and corporations that are usually owned by non-Indigenous and nonlocal individuals, and it creates a great economy for outsiders more than it does for local Indigenous communities. These examples of El Mirador and Hierve el Agua demonstrate the importance of placing Indigenous communities at the forefront of these conversations around tourism. It also demonstrates the importance of letting Indigenous communities manage and govern their lands as opposed to letting non-Indigenous peoples continue to manage and determine how Indigenous lands should be governed and managed.

Land Back

Throughout the Americas we continue advocating for "land back." Many have conflated this as a means for massive deportation, displacing everyone who is not Indigenous to these lands. However, land back means returning the autonomy and right to manage and steward our landscapes back to the Indigenous peoples and communities who were displaced from their lands due to settler colonialism.[20] We have to constantly continue fighting to even be considered, included, or consulted when settler entities make decisions that impact the management and care of our lands. This should not have to be the case as our autonomy to manage and care for our lands should be returned to us. As Indigenous peoples we should not be afterthoughts or continue to be footnotes in work, practices, and policies that govern how our natural resources are allocated and managed. Many of these decisions continue to be extractive, thus accelerating climate change instead of curbing it. Land back allows us to reclaim our self-autonomy and determination, just like the Zapatistas did in 1994.

We should be able to completely depend on our lands to provide education to the children in our communities, lessen the dependence on industrialized food practices and corporations, and sustain our own economies. This will allow us to continue our self-sufficiency as government entities never provide any support or aid to our communities. Land back is what fuels many of our resistance movements that require us to litigate and fight against oppressive structures that continue to govern our lands. This also allows us to practice our ceremonies at our ancestral and sacred sites, as this is oftentimes denied to our people. Due to tourism, some of our sacred and ancestral sites have been opened to the general public, making it difficult for us to even have the opportunity to practice our own ceremonies privately and only with our community members.

Land back is also a movement that embodies deconstructing the notions of what sites are considered "ancient ruins." Describing our sacred sites as ancient ruins dismisses the existence of the descendants of those ancestors who were able to build these great infrastructures around Latin America. People keep forgetting that our communities that are descendants of these heavily romanticized civilizations are still living under extreme poverty, health disparities, and political turmoil in what is now known as Southern Mexico and Central America. The reference to our ancestors as ancient civilizations while not acknowledging that the descendants of these same civilizations continue to fight for their resistance and land liberation deems our history as something only rooted in the past and not in the present. When speaking about ancient ruins and civilizations, it is important to acknowledge our existence. Colonialism destroyed many of our knowledge systems due to the illnesses it introduced to our lands and the extractive practices it continues to maintain. Any land loss is a cultural loss. Continuing this conversation around ancient ruins and civilizations continues to dismiss our resistance and resilience that allows us to continue adapting to modern times and technology. Our relative Rigoberta Menchú, a Maya K'iche' who was awarded the Nobel Peace Prize in 1992, worded this beautifully:

> We are not myths of the past, ruins in the jungle, or zoos. We are
> people and we want to be respected, not to be victims of intolerance
> and racism.

In my Maya Ch'orti' and Zapotec communities, I see how archaeology and anthropology continue to perpetuate this narrative that we are things of the past. These are the fields that tried to trap us behind glass museum cases while supporting colonial endeavors to colonize and conquer our lands. Our histories were heavily documented by anthropologists because they thought we were going to go extinct, and these new cultures they came across were deemed as savage, wild, and mystical. Land back also equips us to reclaim our narratives, histories, and most important our sacred sites that have been converted into tourist attractions or are in the midst of becoming attractions.

Land back will also reduce the extreme displacement Indigenous peoples of Latin America continue to face. For example, Indigenous communities of Central America are forced to relocate to the United States in search of a better life. However, immigration policies that were passed in 2018 created a zero-tolerance policy that asserts that anyone trying to cross the border will be prosecuted under the Department of Justice.[21] Unfortunately this policy also separated children at the border, many Indigenous children from Central America. Juanita Cabrera López, Maya Mam, Executive Director of the Mayan League, has mentioned how criminalizing immigration and family separation policies impact Indigenous Maya children. She states,

> Since December 2018, there have been at least five Maya children and one young Maya woman who have died at the United States (U.S.)-Mexico border under U.S. custody, or killed by Federal officials; Claudia Patricia Gómez González (Maya Mam Nation, 20 years old), was shot in the head by a Customs and Border Patrol agent in Texas on May 23, 2018, after crossing the border. Jakelin Caal Maquin (Maya Q'eqchi' Nation, 7 years old), died of a bacterial infection on December 8, 2018. Felipe Gómez Alonzo (Maya Chuj Nation, 8 years old) died on Christmas Eve 2018 of flu complications. Juan de León Gutiérrez (Maya Ch'orti' Nation, 16 years old), died on April 30, 2019, from a brain infection caused by an untreated sinus infection. Wilmer Josué Ramírez Vásquez (Maya Ch'orti' Nation, 2½ years old), died of pneumonia on May 16, 2019. Carlos Gregorio Hernández Vásquez (Maya Achi Nation, 16 years old) died on May 20, 2019, after an influenza A diagnosis.[22]

I have deep respect for the work Juanita and the Mayan League are doing because this is really heartwork. It hurts to step back and realize that Indigenous peoples are not just dying or facing violence when they protect their Indigenous lands but are also facing the same impacts when they choose to leave and flee in search of a better life. Not only are adults suffering, Indigenous children are facing death and separation from their families at the border.[23] For many this is news they get to read, but for many Central Americans, in particular Indigenous peoples from Central America, these stories are about our families who are at the border separated under the policies the United States continues to uphold that criminalize immigration.

In 2020 I was informed that my cousin, his wife, and two children were making their journey to the United States. They were captured on their way here and the two children were separated from their parents. It hurts knowing that my direct family and community members have to endure such pain after risking their lives in search of a better life. Healing Indigenous landscapes means not only healing our natural resources (both abiotic—nonliving—and biotic living). It means making room so that Indigenous peoples can also heal from everything they have to experience every single day. Those of us with some power and privilege, especially Indigenous women, tend to spread ourselves too thin so that we can provide support and aid to our communities, sometimes failing to put ourselves, our needs, or even our health first. Sometimes this is impossible because there is always someone from our communities who needs help.

In 2020 not only was I surviving the pandemic like everyone else, I was also mobilizing and organizing to support my people during the earthquake that occurred in Oaxaca and the hurricanes that impacted Central American countries, and supporting families who were separated at the border, including my own, and other Indigenous communities who I have close relationships and kinships with that were facing food insecurity. As I mentioned in this book, we face the highest ecological debt, yet those who are responsible for what the climate change our planet is undergoing continue to profit and benefit from settler colonialism.

Healing in the Diaspora

I sat down with Juanita Cabrera López to highlight the work that she and Mayan League, the organization she leads, together with many community members, are doing for Indigenous peoples in the diaspora, especially in regard to immigration policies. These immigration policies have resulted in the death of many of our Indigenous Maya children and the forced sterilization of Indigenous women.[24] I believe that in order to heal Indigenous landscapes, we must also center the voices of Indigenous peoples who have been forced into displacement because of what they are facing back in their ancestral lands, Mexico, Central America, and South America. I believe that we must bring to the light and forefront all Indigenous voices from throughout the Americas to heal as many of us, like myself, are now displaced.

Can you tell me more about Mayan League? What is your organization's mission and how does this mission connect to the environment?

The Mayan League was born out of conflict, and it was born during the genocide and the internal armed conflict that led to our community's forced displacement. Those who were being targeted at that time were many of our elders and community members who were standing up for our rights to our lands, to be treated as human beings, to be respected, especially because we are Indigenous and because we're Maya. More than two hundred thousand of our people were killed or disappeared back in our ancestral homelands, and many of our elders and of our community leaders and organizers were the ones that were targeted. So Mayan League became a movement in [the diaspora]. Our core work is to celebrate our culture and our history and our identity as Indigenous peoples while acknowledging that we live in contemporary times. Despite modernization, we always bring our ancestral knowledge and traditions with us and through this unity between our peoples, we can also unite with other Indigenous people and non-Indigenous people who are in solidarity with human rights, who respect

Mother Earth, and respect our different worldviews. At the end, it's for the benefit of not just the Maya people, but really all of humanity, because this is our common home. And if each of us doesn't do our part, we're all affected. And so making sure that our traditions in our ancestral ways are a key part of all of the work that we do in the Mayan League to stay connected to our founding history as an organization. And then today we still see those initial principles of why we came together. To continue being present as the number of our peoples who continue to be forced out of our ancestral lands increases.

How is the "border crisis" an environmental and climate justice issue?

The border crisis is a process of colonization and the imposition of borders that has been occurring for more than five hundred years. With the creation of nations and states that were violently placed on our ancestral lands through slavery, through killing, through rape, through these very violent processes that never recognized us and only saw our lands and us as commodities, we have been forced to migrate. And so when you have a model already settled over our ancestral lands and territories, and not just the Maya, but all across the *Abya Yala*, and really across the world and how [this] imposition of Western-based models and colonial structures continue to degrade our natural resources to the point that there is no more resources for us Indigenous peoples, original stewards of the lands, we are forced to migrate. What we continue to see today [though the climate justice issue our communities are facing] is a legacy of colonization. What happened with the Maya communities was the model of the government and the state that was created to further policies and create a plan for a small elite, settlers and white people. So everything that was created was to perpetuate control of economic structures, of land, of wealth, within a small minority that were not from our lands originally [settlers]. Even today, there is a lack of legal recognition of our lands and territories, or the very best lands are in the hands of a minority. What that ends up meaning is that when people ask, if climate change is happening, why can't Indigenous people start going to traditional agriculture or traditional ways? My response has been, on what land? When all of our lands have been taken from us, where are we supposed to go? I think there's a lack of awareness of the intrinsic way in which Indigenous peoples and our particular relationship with our lands and territories are not understood. And it is because of the way that we have cared for Mother Earth and Mother Nature, that it is no secret that the largest biodiversity across the world is found on Indigenous people's lands. This is the way of how we are in relationship with our environment. So what has happened is that transnational corporations have come in because they want the wealth of what's under our soil. This is why our displacement is a climate justice issue impacting Indigenous peoples, not just an immigration crisis issue.

What is something you wished people in the United States understood more when it comes to the forced displacement Indigenous peoples from Central America are facing?

I think there is a tremendous amount of historical amnesia, people don't understand history, I think a lot of people think that we [Indigenous peoples from south of the border] just showed up. That's one misconception. Another one is that we all want to be here and it's the American dream. The reality is, if you ask a lot of our Indigenous peoples whether they wanted to leave their ancestral lands, they'll say, we never wanted to leave. Home is not here. There are over five hundred years of displacement and genocide and theft that is embedded in our story and trauma. Forced displacement is not just from 2014 or the 1980s. It is really a history of hundreds of years of being consistently and constantly displaced for what we have and what we've carried. Whether it's our land or whether it's theft of our culture, or whether it's theft of our languages, of our traditions. There's a high level of past aggressions against Indigenous peoples that are now coupled with the lack of recognition that we are Indigenous; this makes us more vulnerable. In our countries of origin, what a lot of people in the United States do not know, is that our Indigeneity as Maya people does not appear or is reflected in our birth certificates. It does not appear in our IDs. And it becomes a process of statistical omission, and even today, the [government would say] that we are about 40 percent of the national population in Guatemala when we know we're probably really closer to 80 percent, but we have no official data backing that up. It is really in the interest of states to deny our existence and to deny our identity. And we see that happening again, in the United States. Now, as soon as our people cross the border or are at the border, there are two things assumed: one, we are all from Mexico and two, we are all Spanish speakers. And so in the cases happening right now, because our identity as Indigenous peoples has been consistently denied [in our ancestral lands and in the United States], our languages are also being denied that has resulted in harsh consequences. We are experiencing human rights violations for our children, for the families that are coming through, for the women and girls. When it comes to not being able to share if there are medical needs, or if they at risk for sexual violence and rape, or human trafficking, how are you going to get help if access to our Indigenous languages is being denied through the journey to make it to this border? This is a lot of the work we do at Mayan League, trying to provide language services and access to the Indigenous Maya people who are being displaced because of political and climate turmoil.

The work Juanita does alongside Mayan League is an environmental, climate, and food justice project that is coupled with racial justice. She recounts how in Guatemala, similar to the rest of Latin America, there is a lack of recognition of Indigeneity because each nation state operates under the premises of racial caste systems. As I have mentioned, these racial caste systems created hierarchies that make white or mestizos (*ladinos* in Guatemala) powerful and privileged. We see this in politics, where there has not been an Indigenous president in Mexico since Benito Juárez. We see this in land ownership, where Indigenous peoples have to constantly fight for their lands despite large corporations owning most of it and where our identities as Indigenous peoples are a threat to settler states because they do not want to grant us rights as Indigenous peoples. Yes, Mexico ratified its constitution in 2001 to incorporate Indigenous rights, but those rights are still violated. This is denying us our Indigeneities, because for our identities, our environments and the healing of our landscapes are crucial. It is crucial for all of us as Juanita mentioned, because what impacts our Mother Earth will impact all of us, and it is already impacting Indigenous peoples.

Yet, when we are within the United States context, our Indigeneities and rights are further denied because, as Juanita recounts, we are all Latino, Latinx, or Hispanic. This is not to say that all Latinos, Latinx, or Hispanics are Indigenous, but for those of us with this identity, it is imperative that it is recognized. My father always told me that giving us an umbrella ethnic term like *Latino* was to further assimilate those of us Indigenous peoples who are displaced. In the environmental sciences we have moved toward the incorporation of *land acknowledgments*, acknowledging whose Native or

Indigenous lands you are currently on, but there are oftentimes no actions taken to support the same Indigenous communities whose lands are being recognized. As Indigenous peoples, we constantly witness performative allyship, and for those of us who are transnational Indigenous peoples, we witness our erasure or exclusion from Indigeneity discourses around our forced displacement.

As I mentioned before, my father would not have left his lands either. My mother would have wanted to stay in her lands as well to be close to community and family. However, behind their stories is a forced displacement because of the genocide, land theft, and climate change Indigenous peoples of Latin America continue to face. This is why I believe that in the environmental sciences, we must move toward decolonizing not just ourselves but the contexts of Indigeneity we continue to uphold. It is not just Guatemalans, Salvadorans, and Hondurans who are being forced to flee their countries and are traveling in caravans. It is not just Guatemalan, Salvadoran, and Honduran children who have died at the border. It is Indigenous peoples, mostly from Maya nations and pueblos, and as Juanita mentioned, the work of the Mayan League is to assist them so that we can reduce the number of deaths among our Indigenous Maya children, eliminate forced sterilization among our Indigenous Maya women, and prevent family separation of Maya families whose children have also been placed under adoption.[25] Yes, our Indigenous cultures, traditions, and practices are beautiful, but as a community, we must move past this romanticization and seek action, support, and respect through decolonization.

> *When we say healing our Indigenous landscapes, what does that mean to the Indigenous peoples from Central America being forced to flee their ancestral lands?*
>
> Healing incorporates the critical aspect of reconnecting with who we are, especially for Indigenous peoples. A lot of times people will ask if you have been forced to leave somewhere,

how is it that you continue your permanence with your place of origin or your traditions, or your ceremonies, if you can't be in those spaces? Unfortunately, the reality is that many of us were forced out, but that doesn't mean that we stop being Indigenous when we've been forced to leave. We carry those teachings with us and hopefully some of us have been blessed to have elders near us or the teachings from our own families, from our own parents. What I'm seeing now are the seeds of this displacement. Those of us who grew up in settings of displacement from when we were little or those of us who were born outside of our ancestral lands. Healing is really about reconnecting with and becoming connected with the land, wherever you are. Having respect for how to care for the land and our environment and giving things. And it doesn't necessarily have to be in all the ways that the proper ancestral ceremonies are because maybe we can't do that, but we put our tobacco down or we offer a prayer by the water, or we have a place in our home for an altar and reflection. I think there's a lot of process right now in taking away the layers of self-preservation [decolonizing], where we have publicly denied our identity because we have often been killed because of our identity.

Decolonizing: Peeling Onions

I often explain decolonization to my students using the metaphor of peeling onions. It is important to mention that colonialism introduced many layers that need to be dismantled, so in order to truly reach decolonization, all these layers must be dismantled. This means that like onions there are multiple layers embedded in ourselves and our systems that need to be peeled off. As we decolonize these layers, we are peeling layers off from this onion until we can get to the core, which is when we truly reach decolonization. Now when it comes to decolonizing larger systems or entities such as universities, we are talking about sacks of onions. This means that at larger entities and institutions, decolonizing means undoing more layers and systems embedded in these layers. Yes, it takes longer to peel a sack of onions than it does one onion. Like peeling onions, we will shed tears, especially Indigenous peoples who have to deal with settler colonialism that started with our genocide and, in some cases, an ongoing genocide that is taking place against our people. Getting to the core of decolonization means that we have dismantled systems and layers that are deeply rooted in our societies. It is a long journey and oftentimes a lifelong journey.

I strongly believe that in Indigenous discourses, the Black community needs to be integrated, in particular of those with African American heritage and lineage, because we often hear the statement "Stolen land, stolen people"[26] without understanding the Indigeneity that both comparisons share. This statement refers to stolen Indigenous lands and stolen Indigenous peoples. The Indigeneity of Black people was stolen from them when they were forcefully displaced from their ancestral lands into the Americas because of slavery. Yes, I have gotten in many arguments, even with Indigenous community members and colleagues, because they do not believe Indigeneity discourses of the Americas should include Black people, but to me it does. Given my positionality as a transnational Indigenous woman from the Americas, I can see through my lived experience how assimilation lures us when we are displaced. Now imagine being displaced overseas in foreign lands and being separated from your teachers, healers, mothers, and knowledge keepers. Slavery stole the Indigeneity of the Black community, and Indigenous peoples, without negating the existence and identity of Black Natives or Afro-Indigenous peoples, must dismantle the anti-Blackness that exists in our communities and embrace the Black community in our Indigeneity discourses. I can only think of how our Indigenous children who have been placed under adoption will also not have access to their healers, teachers, and mothers, who will allow them to sustain their Indigeneity.[27] Healing hurts, and everyone experiences healing differently, but as Indigenous peoples we must make space for all Indigenous peoples, whether they are from the Americas or are descendants of slavery, to heal and join us in our healing circles.

I personally believe Black people (descendants of slavery) in the Americas are neither settlers or allies but are Indigenous peoples who were forcefully displaced from their ancestral lands because of slavery.[28] They were enslaved because they already held to that ancestral knowledge of caring and stewarding the land, and I will continue to include them in my discourses around Indigeneity despite the opposition I will continue to receive, even from other Indigenous colleagues and community members. This, to me, is a layer I have to decolonize in my communities as I believe that in order to dismantle that layer of anti-Blackness in our communities, we must understand that the slavery discourse fails to highlight and explain the Indigeneity of Black people and that Black people have to reclaim their Indigeneity in order to heal. As Indigenous people of the Americas, we must make room for this healing.

Indigenous peoples, especially those of us who have been displaced, must undergo a hard journey of healing because displacement, especially forced displacement, is not something we can heal from easily. This also means many of us have to reclaim our Indigenous languages and celebrate our identities no matter where we are because, as Juanita reminded us, we carry those traditions and cultures with us. We have been forced to forget our ancestral knowledge because it was violently taken from us, burned, or stolen. For transnational Indigenous peoples, part of healing also incorporates leading people into their own self-reflection, because unfortunately our brothers and sisters in the Latino movement oftentimes carry that racism, and Black and Indigenous people are still oppressed within this status and structure that we have been placed in due to the racial hierarchies and caste systems. How Latin American countries were also created through genocide and slavery needs to be acknowledged as well.

These tend to be very uncomfortable conversations for those who benefit from whiteness in Latin America, and for us Indigenous peoples as well, because when we question the dominant narratives and movements, we receive a lot of pushback and witness a lot of white fragility and guilt that is projected on those most marginalized under whiteness in the United States and Latin America. There are levels of oppression within the Latino discourse and movement that objectify Indigenous peoples and Indigenous women because of our beautiful clothing and how we celebrate who we are. Settler colonialism taught us that as Indigenous peoples we were not created equal despite the Americas being our ancestral homelands. In the environmental justice movement, we must reflect on who are the voices left out from platforms, from the advocacy we continue to lead, and the recognition that is not granted to those who have been doing this kind of work for generations.

This is why sometimes for us, Indigenous peoples, decolonization seems unrealistic given that we continue to be afterthoughts in many environmental policies, projects, and initiatives. We are the research subjects but rarely the researchers. We continue to be given roles that continue to take our autonomy from us as opposed to letting us lead and steward our environments. Our knowledge systems continue to be dismissed and their validity questioned. I still recall the first time I wrote about the knowledge my elders had passed to me about our fisheries for one of my graduate courses and the professor asked me, "Is this Jessica's theory? Where is your citation? You need to cite everything you write about." This has stayed with me as I continued my graduate career and journey as there is little Indigenous literature about our ways of knowing written by us. Most of the scholarship and literature is written about us and not by us. Writing this book has allowed me to dismantle this layer that was built around me during graduate school. Like an onion planted on the ground, a new layer grew on me, and this was the layer of needing to cite our lived experience, ways of knowing, and my life experience.

Through writing this book, I am able to dismantle that layer as I am finally able to write a book about my life, my communities, and uplift the voices of those of us who are silenced, ignored, and dismissed. I hope this book can help Indigenous scholars, community members, and our relatives see themselves as scientists. I believe every Indigenous person carries their own scientific knowledge; for some of us it just means that we have to reclaim our knowledge that has been lost because of settler colonialism and how it impacted us individually.

Indigenous peoples do not need college degrees for our knowledge to be accepted or validated. It is time we stop trying to have our knowledge validated and rather build our own tables and be the ones who validate Western knowledge systems. Our knowledge has been present in the Americas since time immemorial, so we are not the ones who need our knowledge, which has been formulated since then, to be validated. Decolonization is a movement Indigenous peoples must lead in the Americas as these are our lands, and this means that the layers of settler colonialism must be dismantled through our leadership. Decolonization is something allies can support, but not co-opt. This is an ongoing effort that will hurt and make us cry, but like peeling onions, our tears will help us heal. The tears we shed will be the tears from our ancestors as they endured harder times and many of them gave the ultimate sacrifice—their lives— to ensure our cultures did not disappear. My grandmother always told me that when she left this world, my tears would be her way of speaking to me, because this is how our ancestors communicate with us. Since tears have healing properties, which Western science has also concluded to be true,[29] our ancestors communicate through us in our tears because when we are healing, they are also healing.

As I continue to work as an Indigenous environmental scientist and advocate, I will embrace the tears I shed and not see them as a sign of weakness. My tears are my grandmother and my ancestors communicating through me. This is why I know writing this book was a form of healing for my relatives, for my community members who provided their testimonies, and for me, because I had never embraced or experienced so much crying in completing a task like writing this book was able to provide. When we heal ourselves we heal landscapes, and it is time we create space for Indigenous peoples to heal as we move forward in life. Community makes space and room for us to grow, continue learning, and embrace our flaws. This is part of the healing process we have to embrace and make room for.

PHOTO 8.1: Ana's grandmother. Photographer: Ana Laura Palacios Cepeda

NOTES

Introduction

1   Roddy Brett, *The Origins and Dynamics of Genocide: Political Violence in Guatemala*, Springer, 2016; Vanessa Jaramillo-Cano, "An Examination of the Varying Role of the United Nations in the Civil Wars of Rwanda and El Salvador" (honors college thesis, University of Nevada Las Vegas, 2012), https://digitalscholarship.unlv.edu /cgi/viewcontent.cgi?article=1056&context=award.

Chapter 1: Indigenous Teaching: Nature Protects You as Long as You Protect Nature

1   Knut Walter and Philip J. Williams, "The Military and Democratization in El Salvador." *Journal of Interamerican Studies and World Affairs* 35, 1 (1993): 39–88.

2   William Stanley, *The Protection Racket State: Elite Politics, Military Extortion, and Civil War in El Salvador* (Temple University Press, 2010).

3   Ignacio Martín-Baró, "De la guerra sucia a la guerra psicológica: el caso de El Salvador." Psicología social de la guerra, El Salvador: UCA (1990).

4   Reagan Martin, *The Martyr of El Salvador: The Assassination of Óscar Romero*, Vol. 2 (Absolute Crime, 2020).

5   Grupo Parlacen and Asamblea Legislativa, "Frente Farabundo Martí para la Liberación Nacional."

6   Mac Chapin, "La población indígena de El Salvador," Mesoamérica 12, 21 (1991): 1–40.

7   Wather René Molina, "Censura previa: ¿reducción a la libertad de prensa? El Salvador, durante el régimen de Pío Romero Bosque, 1927–1929," *Revista de Humanidades y Ciencias Sociales*, (5), 65–111, (2013); Jorge Barraza Ibarra, *Historia de la economía de la provincia del Salvador desde el siglo XVI hasta nuestros días*, Universidad Tecnológica de El Salvador, 2005.

8   Mariella Hernández Moncada, "Pueblos Indígenas de El Salvador: La visión de los invisibles," (2016).

9   Robin Maria DeLugan, "Commemorating from the Margins of the Nation: El Salvador 1932, Indigeneity, and Transnational Belonging," *Anthropological Quarterly* (2013): 965–994.

10 Victor Bulmer-Thomas, *The Political Economy of Central America since 1920*, Cambridge University Press, 1987, https://doi.org/10.1017/CBO9780511572029.

11 Reginald Horsman, *Race and Manifest Destiny* (Harvard University Press, 1981).

12 Engracia Loyo, "La empresa redentora. La casa del estudiante indígena," *Historia Mexicana* (1996): 99–131.

13 Alexander S. Dawson, "'Wild Indians,' 'Mexican Gentlemen,' and the Lessons Learned in the Casa del Estudiante Indigena, 1926–1932," *The Americas* (2001): 329–361.

Chapter 2: Ecocolonialism of Indigenous Landscapes

1   George Yancy, "Feminism and the Subtext of Whiteness: Black Women's Experiences as a Site of Identity Formation and Contestation of Whiteness," *Western Journal of Black Studies* 24.3 (2000): 156.

2   Ula Taylor, "The Historical Evolution of Black Feminist Theory and Praxis," *Journal of Black Studies* 29.2 (1998): 234–253.

3   La Verne Madigan, *The American Indian Relocation Program* (The Association, 1956).

4   Daniel R. Wildcat, *Red Alert!: Saving the Planet with Indigenous Knowledge* (ReadHowYouWant.com, 2010).

5   Matt Driscoll, "The Invisibles: Seattle's Native Americans," *Tulalip News*, 11 Dec. 2014, www.tulalipnews.com/wp/2014/12/11 /the-invisibles-seattles-native-americans-2/.

6   Ben Vinson III, *Before Mestizaje: The Frontiers of Race and Caste in Colonial Mexico*, Vol. 105 (Cambridge University Press, 2017).

7   Patricia Seed, "Social Dimensions of Race: Mexico City, 1753," *Hispanic American Historical Review* 62.4 (1982): 569–606.

8   Alexander Scott Dawson, *Indian and Nation in Revolutionary Mexico* (University of Arizona Press, 2004).

9   Hubert C. De Grammont and Horacio Mackinlay, "Campesino and Indigenous Social Organizations Facing Democratic Transition in Mexico, 1938–2006," *Latin American Perspectives* 36.4 (2009): 21–40.

10 Monika Glowacki et al., "News and Political Information Consumption in Mexico: Mapping the 2018 Mexican Presidential Election on Twitter and Facebook," The Computational Propaganda Project (2018).

11 Fernando Benítez, *Un indio zapoteco llamado Benito Juárez* (Punto de lectura, 2007).

12 *An Issue of Sovereignty*, www.ncsl.org/research/state-tribal-institute/an-issue-of -sovereignty.aspx.

13 Heather J. Shotton, "Beyond Reservations," *Measuring Race: Why Disaggregating Data Matters for Addressing Educational Inequality* (2020): 119.

14 Jessica Hernandez, *Indigenous Lands before Urban Parks: Indigenizing Restoration at Discovery Park, WA*, (Diss., University of Washington, 2020).

15 "Yes on 1631," The Nature Conservancy in Washington, www.washingtonnature .org/yes-on-1631.

16 Saiwa Conejo, "A Blunt Stakeholder Analysis on Initiative 1631 and How It Failed at the Ballot," (2019).

17 Guillermo De la Peña, "A New Mexican Nationalism? Indigenous Rights, Consti- tutional Reform and the Conflicting Meanings of Multiculturalism," *Nations and Nationalism* 12.2 (2006): 279–302.

18 Francisco López Bárcenas, *Autonomía y derechos indígenas en México*, Vol. 10 (Unam, 2005).

19 Suprema Corte de Justicia de la Nación México, José Antonio Caballero Juárez, y Laura Martín del Campo Steta, Political Constitution of the United Mexican States, Poder Judicial de la Nación, Suprema Corte de Justicia de la Nación, 2010.

20 Edgar F. Love, "Legal Restrictions on Afro-Indian Relations in Colonial Mexico," *The Journal of Negro History* 55.2 (1970): 131–139.

21 Felipe Gomez, "Perspectivas de la lucha indígena en México: el Congreso Nacional Indígena y los zapatistas," [2017] Congreso Internacional de Ciencias Sociales, 2017.

22 "¿Qué Es el CNI?" *Congreso Nacional Indígena*, 9 Jan. 2018, www.congresonaciona lindigena.org/que-es-el-cni/.

23 Natalia Rentería Nieto and Gabriel Alejandro Diez Sánchez, "El papel de la mujer dentro de la propuesta de la candidatura indígena de gobierno en México 2018," Dykinson eBook (2019): 59.

24 Marie-Monique Robin, Jeffrey M. Smith, and Mosie Lasagna, *The World According to Monsanto* (Image & Compagnie, 2008).

25 Karen Hudlet, "Maya Beekeepers Stand Up to Monsanto's Genetically Modified Soy," *Fighting the Tide* (2017).

26 Mauricio Feliciano López Barreto, "La decolonialidad como alternativa para la conservación de la biodiversidad. El caso de la meliponicultura en la Península de Yucatán," *Península* 16.1 (2021): 29–53.

27 Laura Huicochea Gómez, "Dulce manjar: sabores, saberes y rituales curativos en torno a la miel de las meliponas," *Ecofronteras* (2011): 22–25.

28 Rogel Villanueva-G, David W. Roubik, and Wilberto Colli-Ucán, "Extinction of Melipona Beecheii and Traditional Beekeeping in the Yucatán Peninsula," *Bee World* 86, no. 2 (2005): 35–41.

29   Ivan P. Novotny et al., "The Importance of the Traditional Milpa in Food Security and Nutritional Self-Sufficiency in the Highlands of Oaxaca, Mexico," *PLOS ONE* 16.2 (2021): e0246281.

Chapter 3: Birth of Western Conservation

1   National Geographic Society, "Conservation," *National Geographic Society*, 5 June 2019, www.nationalgeographic.org/encyclopedia/conservation/.

2   Michael E. Marchand and Wendell George, *The River of Life: Sustainable Practices of Native Americans and Indigenous Peoples* (Gruyter/Higher Education Press, 2014).

3   Felipe Meneses Tello, "El desastre de la documentación indígena durante la invasión-conquista española en Mesoamérica," LIS Critique (Library and Information Science Critique): *Journal of the Sciences of Information Recorded in Documents (Crítica Bibliotecológica: Revista de las Ciencias de la Información Documental)* 4.2 (2012): 20–32.

4   Isaac Kantor, "Ethnic Cleansing and America's Creation of National Parks," *Public Land & Resources Law Review* 28 (2007): 41.

5   Marianne George, "Polynesian Navigation and Te Lapa—'The Flashing,'" *Time and Mind* 5.2 (2012): 135–173.

6   U.S. Department of the Interior, "Organic Act of 1916," National Parks Service, www.nps.gov/grba/learn/management/organic-act-of-1916.htm.

7   Marc Leon Rocher, "Informe de la situación forestal de El Salvador y bases para la elaboración de un plan estatal de reforestación," (1951); Howard E. Daugherty, "The Montecristo Cloud-Forest of El Salvador—A Chance for Protection," *Biological Conservation* 5.3 (1973): 227–230.

8   Alejandro Marroquín, "El problema indígena en El Salvador," *América Indígena* 35.4 (1975): 747–771.

9   Josh Trapani and Katherine Hale, "Science and Engineering Indicators," National Science Foundation, ncses.nsf.gov/pubs/nsb20197/demographic-attributes-of-s-e-degree-recipients.

10 Jovana J. Brown, "Treaty Rights: Twenty Years after the Boldt Decision," *Wicazo Sa Review* (1994): 1–16.

11 "University of Washington Human Resources," UW Human Resources, Director for University of Washington Botanical Gardens, http://uwhires.admin.washington.edu/.

Chapter 4: Indigenous Science: Indigenous Stewardship and Management of Lands

1   Maarten Jansen, "The Search for History in Mixtec Codices," *Ancient Mesoamerica*, vol. 1, no. 1 (1990): 99–112.

2   Danny Zborover, "From '1-Eye' to Bruce Byland," *Bridging the Gaps* (University Press of Colorado, 2015), 1.

3   Felipe Meneses Tello, "El desastre de la documentación indígena durante la invasión-conquista española en Mesoamérica," *LIS Critique (Library and Information Science Critique): Journal of the Sciences of Information Recorded in Documents (Crítica Biblio-tecológica: Revista de las Ciencias de la Información Documental)* 4.2 (2012): 20–32.

4   National Geographic Society, "Columbus Makes Landfall in the Caribbean," *National Geographic Society*, 3 Sept. 2014, www.nationalgeographic.org/thisday /oct12/columbus-makes-landfall-caribbean/.

5   M. E. Danubio, "The Decline of the Tainos. Critical Revision of the Demographic-Historical Sources," *International Journal of Anthropology*, vol. 2, no. 3 (1987): 241–245.

6   Tony Castanha, *The Myth of Indigenous Caribbean Extinction: Continuity and Reclamation in Borikén* (Puerto Rico) (Springer, 2010).

7   Bill Mollison et al., *Introduction to Permaculture* (Tyalgum, Australia: Tagari Publications, 1991).

8   Marie-Josée Massicotte and Christopher Kelly-Bisson, "What's Wrong with Permaculture Design Courses? Brazilian Lessons for Agroecological Movement-Building in Canada," *Agriculture and Human Values* 36.3 (2019): 581–594.

9   Scott Mann, "Episode 1223: The Cost of a Permaculture Design Course (Permabyte)," The Permaculture Podcast, 27 September 2012, www.thepermaculturepod cast.com/2012/pdc-cost/.

10 Eve Tuck and K. Wayne Yang, "Decolonization Is Not a Metaphor," *Decolonization: Indigeneity, Education & Society* 1.1 (2012).

11 University of Washington Alumni Association / Alumni Relations, and UW Alumni, "Viewpoint—Fall 2017," *Issuu*, issuu.com/uwalumni/docs/v-fall-2017-high-res/10; Race & Equity Initiative, "New Course Shines Light on Environmental (In)Justice," www.washington.edu/raceequity/2017/04/20/new-course-shines-light-on -environmental-injustice/.

12 Jed O. Kaplan et al., "The Prehistoric and Preindustrial Deforestation of Europe," *Quaternary Science Reviews*, vol. 28, no. 27 (2009): 3016–3034.

13 T. Douglas Price, *Europe's First Farmer* (Cambridge University Press, 2000).

14 Enrique Salmón, "Iwígara: A Rarámuri Cognitive Model of Biodiversity and Its Effects on Land Management," *Biodiversity and Native America* (2000).

15 Enrique Salmón, "Kincentric Ecology: Indigenous Perceptions of the Human–Nature Relationship," *Ecological Applications* 10.5 (2000): 1327–1332.

16 Silvia H. Salas-Morales, Alfredo Saynes-Vásquez, and Leo Schibli, "Flora de la costa de Oaxaca, México: Lista florística de la región de Zimatán," *Botanical Sciences* 72 (2003): 21–58.

17 Elisabeth Malkin, "Growing a Forest, and Harvesting Jobs," *New York Times*, 23 Nov. 2010, www.nytimes.com/2010/11/23/world/americas/23mexico.html.

18 Gerardo Suarez, "Así Conserva Sus Bosques La Comunidad Zapoteca De Santiago Xiacuí—CCMSS," Consejo Civil Mexicano para la Sivilcultura Sotenible, 12 Aug. 2019, www.ccmss.org.mx/desarrollo-forestal-conservacion-bosques-la-comunidad-zapoteca-santiago-xiacui/.

19 Kana Matsuzaki and Bernard Yun Loong Wong, "Forest Management through Community-Based Forest Enterprises in Ixtlán de Juárez, Oaxaca, Mexico," *Sustainable Use of Biological Diversity in Socio-ecological Production Landscapes* 52 (2010): 98.

20 Gleb Raygorodetsky, "Indigenous Peoples Defend Earth's Biodiversity—but They're in Danger," Environment, *National Geographic*, 10 Feb. 2021, www.nationalgeographic.com/environment/article/can-indigenous-land-stewardship-protect-biodiversity-?loggedin=true.

21 Alice Wong, "The Rise and Fall of the Plastic Straw," *Catalyst: Feminism, Theory, Technoscience* vol. 5, no. 1 (2019): 1–12.

Chapter 5: Ecowars: Seeking Environmental Justice

1 "In Memoriam: 28 Indigenous Rights Defenders Murdered in Latin America in 2019," *Cultural Survival*, 28 Jan. 2020, www.culturalsurvival.org/news/memoriam-28-indigenous-rights-defenders-murdered-latin-america-2019.

2 Nani Lakhani, "Mexico's Deadly Toll of Environment and Land Defenders Catalogued in Report," *The Guardian*, 20 Mar. 2020, www.theguardian.com/environment/2020/mar/20/mexico-environment-land-defenders-murdered-rights-indigenous.

3 Dr. Robert Bullard, https://drrobertbullard.com/.

4 Robert D. Bullard and Beverly Wright, "Disastrous Response to Natural and Man-Made Disasters: An Environmental Justice Analysis Twenty-Five Years after Warren County," *UCLA Journal of Environmental. Law and Policy* 26 (2008): 217.

5 N. Gandhi, S. Bhavsar, E. Reiner, T. Chen, D. Morse, G. Arhonditsis, and K. Drouillard, "Evaluation and Interconversion of Various Indicator PCB Schemes for ΣPCB and Dioxin-Like PCB Toxic Equivalent Levels in Fish," *Environmental Science & Technology*, 49(1) (2015): 123–131.

6 Nick Estes, *Our History Is the Future: Standing Rock versus the Dakota Access Pipeline, and the Long Tradition of Indigenous Resistance* (Verso, 2019).

7 Jessica Hernandez and Kristiina A. Vogt, "Environmental Justice in the Pacific Northwest: Developing an Atlas & Website to Identify Indigenous Pillars of Environmental Justice for Policy Recommendations" (thesis, University of Washington, 2017).

8   Jessica Hernandez, "Indigenizing Environmental Justice: Case Studies from the Pacific Northwest," *Environmental Justice* 12.4 (2019): 175–181.

9   Dara Lind, "The 2014 Central American Migrant Crisis," *Vox*, 10 Oct. 2014, www .vox.com/2014/10/10/18088638/child-migrant-crisis-unaccompanied-alien -children-rio-grande-valley-obama-immigration.

10 Heather M. Wurtz, "A Movement in Motion: Collective Mobility and Embodied Practice in the Central American Migrant Caravan," *Mobilities* 15.6 (2020): 930–944.

11 Wendy A. Vogt, "Crossing Mexico: Structural Violence and the Commodification of Undocumented Central American Migrants," *American Ethnologist* 40.4 (2013): 764–780.

12 Cecilia Menjívar and Leisy Abrego, "Legal Violence: Immigration Law and the Lives of Central American Immigrants," *American Journal of Sociology* 117.5 (2012): 1380–1421.

13 David Androff, "The Human Rights of Unaccompanied Minors in the USA from Central America," *Journal of Human Rights and Social Work* 1.2 (2016): 71–77.

14 Randolph Capps, "A Profile of Immigrants in Arkansas," (2007).

15 S. N. McClain et al., "Migration with Dignity: A Case Study on the Livelihood Transition of Marshallese to Springdale, Arkansas," *Journal of International Migration and Integration* (2019): 1–13.

16 Julia Paley, "The Fight to Ban Gold Mining and Save El Salvador's Water Supply," *Foreign Policy in Focus*, 2014.

17 Benjamin Witte-Lebhar, "Water-Strapped El Salvador Suffers Back-to-Back Molasses Spills," *NotiCen: Central American & Caribbean Affairs*, 2016, 1.

18 World Bank Group, "Guatemala's Water Supply, Sanitation, and Hygiene Poverty Diagnostic: Challenges and Opportunities," Washington, D.C.: World Bank, 2018; William Vasquez, "Rural Water Services in Guatemala: A Survey of Institutions and Community Preferences," *Water Policy* 15, no. 2 (2013): 258–268.

19 H. Galindo and J. Molina, "Valoración Estratégica sobre la Importancia del Agua Potable y el Saneamiento Básico para el Desarrollo, la Salud y la Educación en Guatemala," Guatemala City, Guatemala: Red de Agua Potable y Saneamiento de Guatemala, 2007.

20 William Vasquez, "Municipal Water Services in Guatemala: Exploring Official Perceptions," *Water Policy* 13, no. 3 (2011): 362–374.

21 Anne Braghetta, "Drawing the Connection between Malnutrition and Lack of Safe Drinking Water in Guatemala," *Journal AWWA* 98, no. 5 (2006): 97–106, www.jstor .org.offcampus.lib.washington.edu/stable/41312553.

22 Veronica Montes, "Fleeing Home: Notes on the Central American Caravan in Its Transit to Reach the US–Mexico Border," *Latino Studies* 17.4 (2019): 532–539.

23 Assen Kokalov, "Muxes: auténticas, intrépidas, buscadoras del peligro," *Chasqui* 37.1 (2008): 190–193.

24 Gustavo Subero, "Muxeninity and the Institutionalization of a Third Gender Identity in Alejandra Islas's Muxes: auténticas, intrépidas, buscadoras de peligro," *Hispanic Research Journal* 14.2 (2013): 175–193.

25 Marinella Miano Borruso, *Hombre, mujer y muxe'en el Istmo de Tehuantepec* (Plaza y Valdés, 2002).

26 Filiberto Eduardo R. Manrique Molina and Omar Huertas Díaz, "La acción afirmativa como vía de garantía para preservar la tradición del tercer género (muxe') en el Istmo de Tehuantepec, México," *Via Inveniendi et Iudicandi* 15.2 (2020).

27 Diana Manzo, "Impune, asesinato de activista muxe Óscar Cazorla a un año de su muerte," *Aristegui*, 10 Feb. 2020, available at https://aristeguinoticias.com/1002/lomasdestacado/impune-asesinato-de-activista-muxe-oscar-cazorla-a-un-ano-de-su-muerte/.

28 Mark Anderson, "When Afro Becomes (Like) Indigenous: Garifuna and Afro–Indigenous Politics in Honduras," *The Journal of Latin American and Caribbean Anthropology* 12.2 (2007): 384–413.

29 Nani Lakhani, "Fears Growing for Five Indigenous Garifuna Men Abducted in Honduras," *Guardian*, 23 July 2020, www.theguardian.com/global-development/2020/jul/23/garifuna-honduras-abducted-men-land-rights.

30 Martin Mowforth, "Garifuna Community Continues to Suffer Violence," The Violence of Development, 15 Mar. 2021, https://theviolenceofdevelopment.com/garifuna-community-continue-to-suffer-violence/.

Chapter 6: Tierra Madre: Indigenous Women and Ecofeminism

1 Seraina Rohrer, *La India María: Mexploitation and the Films of María Elena Velasco* (University of Texas Press, 2017).

2 Ariel Zatarain Tumbaga, "Indios y Burros: Rethinking "la India María" as Ethnographic Cinema," *Latin American Research Review* 55.4 (2020).

3 Marco Estrada Saavedra, *La comunidad armada rebelde y el EZLN: un estudio histórico y sociológico sobre las bases de apoyo zapatistas en las cañadas tojolabales de la Selva Lacandona (1930–2005)* (El Colegio de México AC, 2016).

4 Chris Gilbreth and Gerardo Otero, "Democratization in Mexico: The Zapatista Uprising and Civil Society," *Latin American Perspectives* 28.4 (2001): 7–29.

5 Terry Wolfwood, "Who Is Comandanta Ramona?" *Third World Resurgence* No. 84, August 1997.

6 Griselda Herrera López and Adriana Morales García, "La presencia de la mujer en los movimientos armados en Chiapas, México, la Comandanta Ramona y el Ejército

Zapatista de Liberación Nacional," *De género y guerra. Nuevos enfoques en los conflictos armados actuales* (2017): 215.

7 Hilary Klein, *Compañeras: Zapatista Women's Stories* (Seven Stories Press, 2015).

8 Eric Rodriguez and Everardo J. Cuevas, "Problematizing Mestizaje," *Composition Studies* 45.2 (2017): 230–272; Dimitri Nesbitt, "Unlearning White Theories of Race."

9 Paul K. Eiss and Joanne Rapport, eds., *The Politics and Performance of Mestizaje in Latin America: Mestizo Acts* (Routledge, 2018).

10 Silvia Marcos, "The Zapatista Women's Revolutionary Law as It Is Lived Today," *Open Democracy* 22 (2014).

11 Laura-Anne Minkoff-Zern, "Pushing the Boundaries of Indigeneity and Agricultural Knowledge: Oaxacan Immigrant Gardening in California," *Agriculture and Human Values* 29.3 (2012): 381–392.

12 Natalia De Marinis, "Rompiendo el silencio: Construcción de Estado y violencia hacia mujeres triquis de Oaxaca, México (1)," *Cosmovisiones: defensa de territorios, empoderamiento femenino e identidad indígena*: 27.

13 "Fourth Anniversary of the Deaths of Bety Cariño and Jiri Jaakkola," AMARC, amarceurope.eu/fourth-anniversary-of-the-deaths-of-bety-carino-and-jiri-jaakkola/.

14 Elena Poniatowska, *Nothing, Nobody: The Voices of the Mexico City Earthquake* (Temple University Press, 2010).

15 Corinne L. Dufka, "The Mexico City Earthquake Disaster," *Social Casework* 69.3 (1988): 162–170.

Chapter 7: Ancestral Foods: Cooking with Fresh Banana Leaves

1 Douglas H. Marin, Turner B. Sutton, and Kenneth R. Barker, "Dissemination of Bananas in Latin America and the Caribbean and Its Relationship to the Occurrence of *Radophouls similis*," *Plant Disease* 82.9 (1998): 964–974.

2 Jim Stout et al., "NAFTA Trade in Fruits and Vegetables," Global Trade Patterns in Fruits and Vegetables. United States Department of Agriculture, Washington, D.C. (2004): 39–51.

3 Kimberly Amadeo, "The World's Largest Trade Zone That Never Happened," The Balance, 28 Jan. 2021, www.thebalance.com/ftaa-agreement-member-countries-pros-and-cons-3305577.

4 Barbara H. Garavaglia, "Central American-Dominican Republic Free Trade Agreement: Sources of Information," (2005).

5 Daniel Workman, "Bananas Exports by Country," World's Top Exports, 14 Feb. 2021, www.worldstopexports.com/bananas-exports-country/.

6 John P. Schmal, "Oaxaca: A Land of Amazing Diversity," Indigenous Mexico, 29 Sept. 2019, https://indigenousmexico.org/oaxaca/oaxaca-a-land-of-amazing-diversity/.

7   Deependra Yadav and S. P. Singh, "Mango: History Origin and Distribution," *Journal of Pharmacognosy and Phytochemistry* 6.6 (2017): 1257–1262.

8   Richard Rhoda and Tony Burton, Geo-Mexico: *The Geography and Dynamics of Modern Mexico* (Sombrero Books, 2010).

9   Ana Claudia Sánchez-Espinosa, José Luis Villarruel-Ordaz, and Luis David Maldonado Bonilla, "Mycoparasitic Antagonism of a *Trichoderma harzianum* Strain Isolated from Banana Plants in Oaxaca, Mexico: Novel Trichoderma Strain Protects against *Fusarium*," *Biotecnia* 23.1 (2021): 127–134.

10 Randy C. Ploetz, "Panama Disease: A Classic and Destructive Disease of Banana," *Plant Health Progress* 1.1 (2000): 10.

11 Stener Ekern, "The Modernizing Bias of Human Rights: Stories of Mass Killings and Genocide in Central America," *Journal of Genocide Research* 12.3–4 (2010): 219–241.

12 Stephen Kinzer, "Efraín Ríos Montt, Guatemalan Dictator Convicted of Genocide, Dies at 91," *New York Times*, 1 Apr. 2018, www.nytimes.com/2018/04/01/obituaries/efrain-rios-montt-guatemala-dead.html.

13 "Maya Ch'orti," Minority Rights Group, 6 Feb. 2021, https://minorityrights.org/minorities/maya-chorti-2/.

14 Paul J. Dosal, *Doing Business with the Dictators: A Political History of United Fruit in Guatemala, 1899–1944* (Rowman & Littlefield Publishers, 1993).

15 Peter Chapman, *Bananas: How the United Fruit Company Shaped the World* (Grove Atlantic, 2014).

16 David A. Graham, "Is the U.S. on the Verge of Becoming a Banana Republic?" *The Atlantic*, 10 Jan. 2013, www.theatlantic.com/politics/archive/2013/01/is-the-us-on-the-verge-of-becoming-a-banana-republic/267048/.

17 Robin Wright and Dhruv Khullar, "Is America Becoming a Banana Republic?" *The New Yorker*, www.newyorker.com/news/our-columnists/is-america-becoming-a-banana-republic.

18 Orrin Henry, *The Complete Works of O. Henry* (Garden City Publishing Company, 1911).

19 Erik Ching, *Stories of Civil War in El Salvador: A Battle over Memory* (UNC Press Books, 2016); Hal Brands, "Crime, Irregular Warfare, and Institutional Failure in Latin America: Guatemala as a Case Study," *Studies in Conflict & Terrorism* 34.3 (2011): 228–247.

20 Dhaval M. Dave, Drew McNichols, and Joseph J. Sabia, "Political Violence, Risk Aversion, and Non-Localized Disease Spread: Evidence from the U.S. Capitol Riot." No. w28410, National Bureau of Economic Research, 2021.

21 Sandra Cuffe, "Hurricane Eta Devastates Central America as U.S. Withdraws from Climate Accord," *The Intercept*, 16 Nov. 2020, https://theintercept.com/2020/11/16/hurricane-eta-central-america/.

22 Ibid.

23 Cristina Estrada and Eva Calvo, "Hurricane Season Arrives Early in Central America," IFRC, www.ifrc.org/es/noticias/noticias/common/hurricane-season-arrives -early-in-central-america/.

24 Nic Wirtz and Kirk Semple, "Guatemala Rescuers Search for Scores of People Buried in Mudslide Caused by Eta," *New York Times*, 7 Nov. 2020, www.nytimes.com /2020/11/07/world/americas/guatemala-mudslide-storm-eta.html.

25 Brent Patterson, "Mayan Q'eqchi' Communities Resist Hydroelectric Dams in Guatemala," Rabble.ca, 11 Feb. 2021, www.rabble.ca/blogs/bloggers/brent-patterson /2019/01/mayan-q'eqchi'-communities-resist-hydroelectric-dams.

26 "Guatemala: Progress in Renace II Hydroelectric Station," CentralAmericaData, www.centralamericadata.com/en/article/home/Guatemala_Progress_in_Renace_II _Hydroelectric_Station.

27 "Peaceful Resistance Cahabón," PBI Guatemala, pbi-guatemala.org/en/who-we -accompany/peaceful-resistance-cahabón.

28 PBI-Canada, "Indigenous Q'eqchi People Demand Justice for Imprisoned Defender Bernardo Caal and Freedom for Rivers from Hydroelectric Dams," PBI USA, https://pbiusa.org/content/indigenous-q'eqchi-people-demand-justice -imprisoned-defender-bernardo-caal-and-freedom.

29 P. Döll and J. Zhang, "Impact of Climate Change on Freshwater Ecosystems: A Global-Scale Analysis of Ecologically Relevant River Flow Alterations," *Hydrology and Earth System Sciences* 14.5 (2010): 783–799.

30 "Bernardo Caal Xol, Defender of the Cahabón River," PBI Guatemala, https://pbi -guatemala.org/en/who-we-accompany/peaceful-resistance-cahabon/bernardo-caal -xol-defender-cahabón-river.

31 Philip L. Munday et al., "Climate Change and the Future for Coral Reef Fishes," *Fish and Fisheries* 9.3 (2008): 261–285.

32 John M. Guinotte and Victoria J. Fabry, "Ocean Acidification and Its Potential Effects on Marine Ecosystems," *Annals of the New York Academy of Sciences* 1134.1 (2008): 320–342.

33 Natalie C. Ban and Alejandro Frid, "Indigenous Peoples' Rights and Marine Protected Areas," *Marine Policy* 87 (2018): 180–185.

34 Fermin Koop, "How Latin America Is Leading the Way for Marine Protection," China Dialogue Ocean, 5 Apr. 2019, https://chinadialogueocean.net/7428-latin -america-leading-marine-protection/.

35 Josh Korman, Michael D. Yard, and Theodore A. Kennedy, "Trends in Rainbow Trout Recruitment, Abundance, Survival, and Growth During a Boom-and-Bust Cycle in a Tailwater Fishery," *Transactions of the American Fisheries Society* 146.5 (2017): 1043–1057.

36 Bryan van Hulst Miranda, "This New Coca Cola Ad Shows Mexico's White Savior Problem," TeleSUR, 27 Nov. 2015, www.telesurenglish.net/analysis/This-New -Coca-Cola-Ad-Shows-Mexicos-White-Savior-Problem-20151127-0003.html.

37   Charles Wilkinson, *Messages from Frank's Landing: A Story of Salmon, Treaties, and the Indian Way* (University of Washington Press, 2006).

38   O. Yale Lewis III, "Treaty Fishing Rights: A Habitat Right as Part of the Trinity of Rights Implied by the Fishing Clause of the Stevens Treaties," *American Indian Law Review* 27 (2002): 281–553.

39   "Treaty of Point No Point, 1855," GOIA, https://goia.wa.gov/tribal-government /treaty-point-no-point-1855.

40   Kent Richards, "The Stevens Treaties of 1854–1855," *Oregon Historical Quarterly* 106, no. 3 (2005): 342–350.

41   Trova Heffernan, *Where the Salmon Run: The Life and Legacy of Bill Frank Jr.* (University of Washington Press, 2017).

42   Charles Wilkinson, *Messages from Frank's Landing: A Story of Salmon, Treaties, and the Indian Way* (University of Washington Press, 2006).

43   Ericka Michal, "Elwha River Dams and Alta Dam: People, Politics, and Ecosystems."

44   Bruce G. Miller, "The Press, the Boldt Decision, and Indian-White Relations," *American Indian Culture and Research Journal* 17.2 (1993): 75–97.

45   Rita Bruun, "The Boldt Decision: Legal Victory, Political Defeat," *Law & Policy* 4.3 (1982): 271–298.

46   Pacific Fishery Management Control, "Treaty Indian Troll Management Alternatives for Ocean Salmon Fisheries—Tribe Proposed 2021," www.pcouncil.org/.

47   Ibid.

48   "State Awards $18 Million in Grants to Recover Salmon," Washington State RCO, 17 Sept. 2020, https://rco.wa.gov/2020/09/17/state-awards-18-million-in-grants-to -recover-salmon/.

49   Courtney Flatt, "Report: Salmon in WA Are 'Teetering on the Brink of Extinction,'" Crosscut, 20 Jan. 2021, https://crosscut.com/environment/2021/01/report-salmon -wa-are-teetering-brink-extinction.

Chapter 8: Indigenizing Conservation: Healing Indigenous Landscapes

1   Francisca De la Maza, "Tourism in Chile's Indigenous Territories: The Impact of Public Policies and Tourism Value of Indigenous Culture," *Latin American and Caribbean Ethnic Studies* 13.1 (2018): 94–111.

2   I. P. Azcona, E. I. J. E. Lugo, A. M. A. Ibarra, and E.B. Baltazar, "Meanings of Conservation in Zapotec Communities of Oaxaca, Mexico," *Conservation & Society*, 18(2), (2020): 172–182.

3   Erika S. Zavaleta, Richard J. Hobbs, and Harold A. Mooney, "Viewing Invasive Species Removal in a Whole-Ecosystem Context," *Trends in Ecology & Evolution* 16.8 (2001): 454–459.

4   Marcy H. Towns, "Where Are the Women of Color? Data on African American, Hispanic, and Native American Faculty in STEM," *Journal of College Science Teaching* 39.4 (2010): 8.

5   Daniel C. Knudsen, ed., *Landscape, Tourism, and Meaning* (Ashgate Publishing, Ltd., 2008).

6   Francisco Bautista, "El Tren Maya: ¿Cuál debería ser el debate?"

7   Adrián Flores, Yannick Deniau, and S. Prieto-Diaz, "El Tren Maya. Un nuevo proyecto de articulación territorial en la Península de Yucatán," en ligne: http://geocomunes.org/Visualizadores/PeninsulaYucatán (2019).

8   Neri Santos Isis Jazmin, "La hegemonía burguesa de la denominada 'cuarta transformación' contra el zapatismo y los pueblos originarios en Resistencia," Memoria de trabajos extensos del III Congreso Mexicano de Sociología.

9   Yolanda Lopez-Maldonado and Fikret Berkes, "Restoring the Environment, Revitalizing the Culture," *Ecology and Society* 22.4 (2017).

10   Ray T. Matheny, ed. "El Mirador, Peten, Guatemala: An Interim Report," No. 45–47 (Provo, Utah: New World Archaeological Foundation, Brigham Young University, 1980).

11   Beatriz García, "The Suspicious Initiative of an Archaeologist to 'Save' an Ancient Mayan City," *AL DÍA News*, 18 June 2020, https://aldianews.com/articles/culture/social/suspicious-initiative-archaeologist-save-ancient-mayan-city/58920.

12   Jeff Abbott, "U.S. Archeologist Seeks to Privatize Maya Historic Sites in the Name of Conservation," NACLA, 27 Aug. 2020, https://nacla.org/guatmala-peten-tourism-hansen.

13   Julia Lindau, "Mayan Ruins in Guatemala Could Become a U.S.-Funded-Tourist Attraction," VICE, www.vice.com/en/article/889qpz/mayan-ruins-in-guatemala-could-become-a-us-funded-tourist-attraction.

14   Ibid.

15   Aura Cumes et al., "Open Letter to Archaeologist, Richard D. Hansen Regarding His Imperialist and Colonial Drive to Expropriate Our Territories and Sacred Sites," TUJAAL.ORG, 2 July 2020, https://tujaal.org/open-letter-to-archaeologist-richard-d-hansen/.

16   "Cierra Hierve El Agua Al Turismo Nacional y Extranjero," Quadratín, 24 Mar. 2021, https://oaxaca.quadratin.com.mx/cierra-hierve-el-agua-al-turismo-nacional-y-extranjero/.

17   "Residents Close Oaxaca Tourist Destination, Claiming Only Outsiders Benefit," *Mexico News Daily*, 25 Mar. 2021, https://mexiconewsdaily.com/news/residents-close-oaxaca-tourist-destination/.

18 "Sala De Justicia Indígena, Un Paso Histórico En México: Adelfo Regino," Instituto Nacional De Los Pueblos Indígenas, www.gob.mx.

19 "Residents Close Oaxaca Tourist Destination," *Mexico News Daily*.

20 Nikki A. Pieratos, Sarah S. Manning, and Nick Tilsen, "Land Back: A Meta Narrative to Help Indigenous People Show Up as Movement Leaders," *Leadership* (2020), doi/10.1177/1742715020976204.

21 "Family Separation under the Trump Administration—a Timeline," Southern Poverty Law Center, 17 June 2020, www.splcenter.org/news/2020/06/17/family -separation-under-trump-administration-timeline.

22 Juanita Cabrera Lopez, Patrisia Gonzales, Blake Gentry, *A Të Qík'xyé Toj Nin K'ul Ex Toj Chg'ajlaj: When We Cross the Mountains and Desert, Indigenous Forced Migration in Abiayala*, www.culturalsurvival.org/publications/cultural-survival-quarterly/te -qikxye-toj-nin-kul-ex-toj-chgajlaj-when-we-cross.

23 Patricia Foxen, *In Search of Providence: Transnational Mayan Identities* (Vanderbilt University Press, 2020).

24 Maya Finoh, "Allegations of Forced Sterilization in ICE Detention Evoke a Long Legacy of Eugenics in the United States," Center for Constitutional Rights, 18 Sept. 2020, ccrjustice.org/home/blog/2020/09/18/allegations-forced-sterilization -ice-detention-evoke-long-legacy-eugenics.

25 Associated Press, "Deported Parents May Lose Kids to Adoption, Investigation Finds," NBCNews.com, 10 Oct. 2018, www.nbcnews.com/news/latino/deported -parents-may-lose-kids-adoption-investigation-finds-n918261.

26 Kathleen A. Brown-Pérez, "From the Guest Editor: An Introduction to Stolen People, Stolen Land, Stolen Identity: Negotiating the Labyrinth of Anglo-American Culture and Law," *Landscapes of Violence* 2.1 (2012): 1.

27 Carmen Monico, Karen S. Rotabi, and Justin Lee, "Forced Child–Family Separations in the Southwestern US Border under the 'Zero-Tolerance' Policy: Preventing Human Rights Violations and Child Abduction into Adoption (Part 1)," *Journal of Human Rights and Social Work* 4.3 (2019): 164–179.

28 Zainab Amadahy and Bonita Lawrence, "Indigenous Peoples and Black People in Canada: Settlers or Allies?" in *Breaching the Colonial Contract* (Springer, Dordrecht, 2009), 105–136.

29 Marjaneh M. Fooladi, "The Healing Effects of Crying," *Holistic Nursing Practice* 19.6 (2005): 248–255.

INDEX

ABOUT THE AUTHOR

JESSICA HERNANDEZ, PhD, is a transnational Indigenous scholar, scientist, and community advocate based in the Pacific Northwest. She has an interdisciplinary academic background ranging from marine sciences to forestry. Her work is grounded in her Indigenous cultures and ways of knowing. Hernandez advocates for climate, energy, and environmental justice through her scientific and community work and strongly believes that Indigenous sciences can heal our Indigenous lands. She is the founder of Piña Soul, SPC, an environmental consulting and artesanias hybrid business that supports Black- and Indigenous-led conservation and environmental projects through community mutual aids and microgrants.

About North Atlantic Books

North Atlantic Books (NAB) is a 501(c)(3) nonprofit publisher committed to a bold exploration of the relationships between mind, body, spirit, culture, and nature. Founded in 1974, NAB aims to nurture a holistic view of the arts, sciences, humanities, and healing. To make a donation or to learn more about our books, authors, events, and newsletter, please visit www.northatlanticbooks.com.